Textbook of Home Science Extension Education

Textbook of Home Science Extension Education

Serene Shekhar (Gote)
and
Santosh Ahlawat

2013
Daya Publishing House®
A Division of
Astral International Pvt. Ltd.
New Delhi – 110 002

ISBN 9789351242062

Published by : **Daya Publishing House®**
A Division of
Astral International Pvt. Ltd. –
ISO 9001:2008 Certified Company –
4760-61/23, Ansari Road, Darya Ganj
New Delhi-110 002
Ph. 011-43549197, 23278134
E-mail: info@astralint.com
Website: www.astralint.com

Laser Typesetting : **Classic Computer Services**
Delhi - 110 035

Printed at : **Chawla Offset Printers**
Delhi - 110 052

PRINTED IN INDIA

Foreword

The extension system is a complex network of various stakeholders, *i.e.*, researchers, extension workers, farmers, NGOs, government and agriculture institutions. The extension system now has been placed in driving seat for rapid transformation of rural life. This book is an elegant piece of work which explains all the basic and applied aspects of home science extension education to the undergraduate students. This book provides a clearly written, well documented text that describes the role of home science extension in raising the overall living standards of rural families by imparting knowledge and skill, offering technical advice, helping to identify problems and opportunities, sharing information, supporting group formation and planning developmental programmes.

The authors through this book gently nurse its readers by simple examples on a vast range of topics that illustrate the practical side of extension work. This book will serve as a useful reference on home science and agriculture extension services to undergraduate students. The illustrative text and references enable the reader to follow the points of interest in detail.

This book entitled *"Textbook of Home Science Extension Education"* by Serene Shekhar (Gote) and Santosh Ahlawat, will contribute significantly by inspiring students for excellent extension work.

I recommend this book for all undergraduate students of home science in SAUs.

Dr. K. Shreedharan
Vice Chancellor,
Sardarkrushinagar Dantiwada Agricultural University
S.K. Nagar, Gujarat

Preface

The purpose of this book is to introduce the basics of home science, its applied branch home science extension and extension system as a whole. Each chapter explains separate aspect of extension education. The book is written on the basis of new syllabus recommended by 4th Deans Committee. The chapters are arranged in order of increasing technical complexity. This book is majorly written for undergraduate and post graduate students and also for the extension agents working in this area.

The first few chapters briefly explains emergence of home science education in India, misconcepts attached to home science education and the scope of home science in present era. It also advocates how home science education help in personal and professional development of all members of family.

The next few chapters deals with details about historical development in extension and how home science extension plays vital role in development of women, family and community as a whole. It also explains various agriculture development and women and child development programmes in India.

The third section of the book explains the different Extension Teaching Methods and teaching aids used while extending information in the village because for an extension worker, it is important to know its concept, use, importance and limitations of each of the extension teaching methods so that they can use it efficiently.

The fourth section is written with the view that without communication, there is no way to exchange thoughts, ideas and feelings. Therefore for extension agent it is important to understand communication techniques, ways to diffuse innovation in society, factors that influence diffusion and consequences associated with adoption of innovation.

The fifth section deals with planning involved in formation of developmental programme, its implementation and evaluation because it is essential for extension professionals to build up programmes for rural development. It helps extension agent to develop plan of action and to set priority in a programme to achieve goal.

It is said extension work can be successful only with skilful extension agents. We hope that, in writing this book, we will, in some way develop skill and understanding in students, for them to become perfect extension worker in future.

Serene Shekhar (Gote)

Santosh Ahlawat

Contents

Instructional Technology

Communication and Diffusion of Innovations

Programme Planning, Implementation and Evaluation

Introduction to Home Science

Chapter-1

History of Home Science

Meaning of Home Science Education

Home Science Education is well structured education of home living. Through its core courses such as clothing and textile, foods and nutrition, human resource development, human development and extension education, it helps to bring economic independence among individual so as to raise the standard of living. Science of Home is concerned with maintenance and enrichment of human relationship within and outside the family, through development and judicious use of all human and material resources to achieve maximal satisfaction for all members of family. Home Science Education advocates for personal and professional development of all members of family.

Philosophy of Home Science Education

Philosophy is an attempt to answer ultimate questions critically. It is comprehensive idea or point of view about anything.

- Ellen Richards, the first President of The American Home Economics Association, in 1909 stated that "Philosophy of Home Science is to achieve freedom of home from dominance of any idea that is against interest of home and society".
- Rajmal Devadas, President of Home Science Association of India described philosophy of Home Science as "Home Science deals with all aspect of life of community and nation." It integrates application of knowledge synthesized from different sciences and humanities to improve human environment, family nutrition, management of resources, child development and consumer competence.

The Contemporary Philosophy of Home Science Education

- Human population must be controlled through Home Science education because unmanageable expansion of family put undue pressure on women so it should be controlled by developing population education material and educating people regarding family planning.
- Home Science Education broadened to relate the nucleus of family to national economy, the national education system and to national political system.
- Inclusion of extension service in Home Science *i.e.,* to make available the knowledge of Home Science to families. Who are devoid of or who are not reached by formal institutions.
- To extend the knowledge of Home Science to villagers through Home Science Training for village

extension workers (Gram Sevikas, Mukhya Sevikas).

- ✰ Encouragement of women to be gainfully employed *i.e.*, to seek and accept paid job, to make them professionally competent.
- ✰ Introduction of Research at postgraduate level to meet needs of society.

Emergence of Home Science Education in India

Home science is systematic study of home. Its aim is utilization of modern science to deal every aspect of home.

- ✰ Home Science education emerged as a field of education in India in early twenties and was referred to as Domestic Science, House Craft or Domestic Economy. It was then offered at primary school level and was not a subject of examination. Its aim was to impart home making skill to the girls to prepare them for future housewives. But then there were neither qualified staff nor equipments or text books.
- ✰ A progressive ruler of Baroda, Maharaja Sayajirao Gaekwad, contributed significantly in promotion of women's education. In 1919 he invited Miss Anna Gilchrist Strong, a home economist from New Zealand to teach domestic science in Maharani's Girls High School in Baroda. The princely state of Baroda was the first to introduce home science at secondary school. Till then home science education was not a subject of examination.
- ✰ For bringing home science education to main stream of education, the first All India Women's

Conference (AIWC) was formed in 1927 under leadership of Mrs. Margaret Cousins. The conference appointed a committee to review the existing education for girls. This committee recommended for suitable education programme for girls, thus the idea of home science as a subject of study was born. In 1929 AIWC founded the All India Women's Education Fund Association under the presidentship of Lady Dorothy Irwin, the then Vicereine and Secretary Mrs. Rustomji. With their efforts funds were collected to start home science college. Hence the Lady Irwin College of Home Science for girls were opened in New Delhi on 11th November 1932.

- ✰ In the year 1936 the Allahabad Agricultural Institute established a department of Home Economics with a two years college course.
- ✰ Later, in 1942 the Women's Christian College and the Queen Mary's College in Chennai jointly started the first degree course in India in Home Science.
- ✰ For the first time, in 1950 - the M.S University of Baroda opened a Faculty of Home Science with Dr. Leena Shah as Dean and Mrs. Hansa Mehta as its Vice Chancellor.
- ✰ In mid 1960-1970's Agricultural University established in most of States and home science education saw a boom in development when the need for its study was recognized as it facilitate students to develop their own resources to achieve optimal and sustainable living.

☆ Today home science education has been established as professional field and is part of schools and college education to cater welfare and development of family, community and nation. It also play significant role in many non-formal and informal educational programmes.

Objectives of Home Science Education

Broadly home science education aims to acquire family happiness, raise ethical standards and improve economic condition through entrepreneurial approach.

☆ To strengthen potentials (skill and knowledge) of home scientists through scientific approaches in home and family living.

☆ To develop managerial skill so that home scientist learns to manage resources to attain goals of individual, family, society and nation.

☆ To help in all round development of home scientists that thereby improve home living.

☆ To prepare for career so as to increase income of family.

☆ To raise standard of living of people by developing entrepreneurial skills in them.

☆ To develop ability of home scientists to manage home and family.

☆ To encourage home scientists for innovative research to meet the challenges in day to day life.

☆ To sensitise home scientists for rural life and its problems.

Characteristic of Home Science

- ☆ It has concern for home and family living
- ☆ It personalizes instructions. It believes that home is place for development of both sexes through equal opportunity and advocates personal and professional development of both sexes within home. It helps to avoid imbalances between roles of men and women.
- ☆ It fuse role of women as home maker dimensions and professional because single diemstional role of women create constraint in all-round development.
- ☆ It advocates better home living by making women/ men economically helping.
- ☆ Home Science promotes spiritual aspect of family living such as moral values of honesty, truth, dependability and objectivity in decision making.

Professional Bodies of Home Science Education at National and International Level

Home Science Association of India (H.S.A.I.)

The Home Science Association of India is national body that was formulated to promote home science education in India.

It was initiated in 1951 at a meeting under the leadership of Smt. Hansa Mehta, the then Vice-Chancellor of the M.S University of Baroda. The Home Scientist who participated was late Mrs. Leela Desai, Dr. Plemmie Kittrel, Dr. Rajammal P. Devadas, Miss Dorothy Pearson, Mrs. Sucykoshy, Dr. (Miss) Florence Theophilus and Miss Nitya Soysa from Ceylon.

Mrs. Dorothy Pearson of Women's Christian College, Chennai, was responsible for forming the constitution which was adopted at the first conference of the Association held at Chennai in 1952.

The basic aims and objectives of the Association are:

1. To promote the welfare of the family through sound home making education.
2. To raise the standards of the profession of the Home Science.

Head quarter of H.S.A.I. is at Vadodara. The basic aims are to be implemented by:

1. Encouraging the establishment and improvement of home making education in schools, training colleges, universities and out of school program.
2. Making available reports, pamphlets and other publications relating to Home Science.
3. Encouraging and aiding research and investigation in the various fields of Home Science.
4. Standardizing and coordinating the course in different institutions.
5. Being instrumental in promoting consumer education for quality products.
6. Undertaking supporting activities for the expansion of Home Science Extension for the benefit of rural home-makers.
7. Developing scientific attitudes and artistic outlook in home making.
8. Exploring the employment opportunities of Home Scientist and raising the professional status of Home Science.

Home Science Association of India was later affiliated to I.F.H.E. (International Federation of Home Economics).

International Federation of Home Economics (I.F.H.E.)

The International Federation for Home Economics is the only worldwide organization concerned with Home Economics and Consumer Studies. It was founded in 1908 to serve as a platform for international exchange within the field of Home Economics. IFHE is an International Non Governmental Organization (INGO), having consultative status with the United Nations (ECOSOC, FAO, UNESCO, and UNICEF) and with the Council of Europe.

Aim of IFHE

- ☆ Provide opportunities for Global networking among professionals
- ☆ Promote the recognition of Home Economics in the everyday lives of individuals and families
- ☆ Promote continuing education in Home Economics
- ☆ Provide opportunities through practice, research and professional sharing that lead to improving the quality of everyday life for individuals, families and households worldwide

Chapter-2

Home Science Education: Misconcepts and Scopes

Home Science education started in 1920s with the objective of making women good homemakers. In due course of time, it has emerged as a professional course that has widened the horizon for women also to become professional Home Scientists. Despite of its varied professional building courses, misconceptions about Home Science studies still exists and even in 21st century it is struggling for its existence. The first approach towards correcting the misconceptions is to make the people aware of all the wrong concepts along with the facts. Few but most common wrong concepts about Home Science education are:

Home Science Education is for Academically Poor Performing Students

Fact

One has to go through tough competitive examination for seeking admission in Home Science especially at Post graduation level in renowned colleges. So the above statement is wrong; instead the statement should be "Home Science education is for competitive students". It is the science that trains students to cope with new developments and growing needs of individuals and face with confidence the new challenges of rapidly changing society. It enables them to attain successful, satisfactory and harmonious living.

Home Science Teaches Merely Cooking, Stitching, Interior Decoration and Child Care

Fact

Home Science includes the aspects like food and nutrition, apparel designing, interior decoration human development and communication management with an emphasis to build scientific knowledge and to develop a temperament for its application in daily and professional life. It demonstrates the scientific facts along with developing different skills among students. Though Home Science teaches cooking, it is based on the science and art of cooking. Home Science deals with principles and methods of cooking for preventing loss of nutrients and adding variety, balance and nourishment in the diet. For example along with developing understanding regarding various nutrients, their requirement and role for different age groups; different nutritionally rich food products are developed under food and nutrition department. These products are developed for ensuring dietary requirement of different groups including children, adolescents, youth, pregnant and lactating women as well as old age people.

Likewise in apparel designing, students learn to identify different kinds of yarns, fabrics, and their formation through off loom weaving techniques, their chemical and physical properties and different treatments for enhancing fabric quality. Additionally, students learn to design apparels suitable for various body structures and for various occasions along with principles and elements of art in relation to clothing and socio-psychological factors affecting clothing choice.

Knowledge of laundry techniques, care of textiles, wardrobe planning, and setup of apparel industry and apparel business management is imparted to students. Further, skill in garment design, garment construction, fashion sketching, creating fashion apparels techniques, dyeing, printing techniques, product development and making pattern using CAD software is developed among students.

In family resource management they become skilled at designing a room for various occasions-formal or informal, for various purposes and age groups. Students learn to manage different human and nonhuman resources, techniques of work simplification and drudgery reduction. Skill in preparation of traditional and modern handicrafts of India is also developed.

Competency in the application of various principles of design, elements of arts and design for the interior and exterior decoration is developed among students. Students get knowledge about space requirement and space designing in residential and institutional buildings and become skilled in preparation of layouts, sketching, drawing, illustrations of floor plans and elevation plans with applications of colors and paints, placement of various types of finishes, fittings and fixtures and management of floor coverings and furnishings. Proficiency in the use of Auto CAD for house designing is also developed.

Under the department of Human and Family Studies, students learn about various developments (physical, social and cognitive) and changes occur in a human since conception till old age. They further learn to deal with numerous developmental problems get sensitized about the needs and problems of children, youths, women, and the aged people under specific situation in the context of changing socio-economic scenario. Skill to design and create play materials for children and planning various kinds of learning experiences through creative activities, planning and implementing programs of parent and the community education is developed. Students are imparted knowledge regarding management of ECCE centers and methods of measuring development and ways of evaluation.

Other important department is Home Science Extension and Communication Management in which field experiences are imparted to the students and technical knowledge of different areas of home science is disseminated to the rural people who are deprived of formal education. It develops skill of script writing for different development media, video film production and evaluation and development of various types of self-instructional print media. It develops competency in computer, photography, and skills in designing and using multi-media modules.

Thus it can be said that Home Science education is a rare combination of scientific knowledge and skill up gradation. It caters to betterment of mankind and improves homemaking skills along with professional skills.

Home Science Studies can be Learned at Home

Fact

No doubt girls can learn so many homemaking skills from their mothers but not necessarily in scientific manner and with scientific reasoning. Secondly now objective of Home Science education is not only to produce good home makers, it has become a professional course. Learning of home science at home will not provide students with professional degree which is necessary for getting job. Thus the subjects of Home Science offered in Home Science institutions supplement students with scientific reasoning that helps them to do things in a more organized manner. Thus it can be said that Home Science enables students to cope up with knowledge explosion and household and related technological advancement that cannot be learned at home.

Home Science Education is Only for Girls

Fact

In today's world there is a predominance of the nuclear family structure and both the partners work hard for earning and for meeting the needs of their family. Women have started to combine their homemaking responsibilities with generating income for the family. Under such conditions, the work load of women is shared by the men and thus both the partners need to be trained to cope-up with home making activities and dealing with various family issues. Hence it is time to change the mindset of people in society and make them to accept the fact that Home Science is beneficial for both girls as well as for boys also.

Studying Home Science Hinders One's Professional Growth

Fact

Home Science education offers to its graduates and post graduates with a wide scope of professional growth *i.e.* for higher studies, government jobs, self employment and entrepreneurship. Home Science students have opportunities to do M. Sc, M. Phil and PhD into five major fields of Home Science. Apart from doing PG and PhD courses in Home Science field, Home Scientists are also eligible for courses in ABM, MBA, Dietetics, NIFT, NID, NIFD, MSW, Food technologies etc. Thus Home Science education widens the horizon for professional growth in varied fields rather restricting it.

Home Science does not Offer Attractive Job Opportunities

Fact

Different areas of Home Science prepare students for a variety of jobs and self employment which ensures economic stability of the family and raise the living standard of family and improve quality of life. Home Scientists can serve as Professors, Research Scientists, Project Officers and Managers in NGOs, Government and Private institutions. The Home Scientists can take the job of Dieticians in hospitals, nursing homes and fitness centers etc or work as Nutrition Counselors and Nutrition Experts in national/ international agencies like, NIN, ICMR, WFP, UNICEF, CARE, CRY, and CHETNA etc. They can work as health and nutrition Journalists, work in consumer forum or can take job as Extension worker in KVK, FTC and adult education/continuing education centers. Having knowledge of ergonomics and work simplification, the home scientists can design ergonomically suitable furniture, farm and household equipments. They can be child counselor and Child development programme officer with specialization in HDFS and FN.

Home Science does not Offer Attractive Entrepreneurial or Educational Opportunities

Fact

This field of study objectively prepares the students to work in any of the following enterprises:

a) Educational Scope

All the graduates from Home Science are eligible for management course like: MBA, Agribusiness Management (ABM). The students with Nutrition as their elective can further opt for Dietitian courses, Textile and Apparel Designing students can opt for NIFT, NIFD, and NID for further studies. The graduates with Extension as specialization can go for MSW and Masters in Agriculture Extension (as few of Agriculture College permits Home Science Extension Graduates for Post Graduate programme in Agriculture Extension). The Graduates with Human Resource Management (HRM) can opt for Masters in Fine Arts; similarly the students with Human Development (HDFS) can for masters in their respective department for further academic development.

b) Entrepreneurial Scope

Knowledge of multidisciplinary fields of Home Science prepares one for various Entrepreneurial Scopes such as:

Department of Food Sciences and Nutrition

This field of study objectively prepares the students to start work as in:

- ☆ Catering services or Canteen centre can be opened for people who are working in factories, offices and do not have time or arrangement to cook meals. Catering service could be provided for parties arranged at home such as kitty parties, birthday parties or anniversary parties etc.
- ☆ Small cottage industry for sale of preserved vegetables and fruits in the form of pickles, jams, jelly, marmalades etc. could be undertaken as profession. Similarly Home Scientist can open Confectionery, Bakery or Ice – cream Parlors.
- ☆ Health Centers for giving guidance to the individuals for keeping fit through diet, exercise and management of obesity and related conditions. Special advice for the dietary needs of people suffering from different diseases can also be provided by the professionals.
- ☆ Small units of ready to cook/ready to serve food could be established to clean or shell the vegetable so as to make these ready for cooking by the house wife. Fresh Fruit chart centre can also be established.

Department of Human Resource Management

Home Scientists can open up centers where they can guide people on following aspects:

- ☆ The Home Scientists can impart training in the art of interior decoration and can also provide services for decoration of various settings like offices, hospitals, school.

- ☆ Hobby Centres could be started for persons interested in learning candle and paper flower making, preparation of decorative articles, soft toys, Rangoli, jewellery designing, pot making, wall painting and making useful articles from the household waste products.
- ☆ Home scientist can open their own guiding center on consumer rights.

Department of Fabric and Apparel Designing

- ☆ Owner of boutique, knitted garment unit, weaving unit and fabric enrichment unit
- ☆ Owner of dry cleaning shop
- ☆ Soft toy making centre

Department of Human Development

- ☆ Child Care/Day Care Center and Mobile Crèches
- ☆ Home Science graduates can run day – care center, crèches, nursery school and after school centers for children of working mothers. It can also be opened in market area so that mothers can leave their children in safe hands and can shop well freely.
- ☆ Old age home can be managed by Home Science graduates.
- ☆ Home Science graduates can open rehabilitation centers for children with impaired senses where they can be taught for developing various employment skills; so that they can earn their living.

Department of Extension and Communication Management

- ☆ Owner of NGO, cyber etc.
- ☆ Video shooting service center
- ☆ Owner of printing unit
- ☆ Professional writer for mass media
- ☆ Multimedia production unit
- ☆ Developer of programmes for media

There are varied ways to enhance image of Home Science in the society as field of professional education. These are as follows:

1. *Comprehension of objectives of the courses to the students*: Efforts should be made to deliver the courses in a way to make students understand the objective of the curriculum. This will help students to explore jobs and entrepreneurial opportunities in related fields.
2. *Getting recognition at state and centre levels*: Recognition should be gained at State and Central level so that general public can be able to know Home Science as an up-coming stream. Getting recognition will also help the Home Scientist to be eligible for various state and national level examinations such as banking etc.
3. *Organizing seminars, symposia, workshops and inviting apex bodies from government and private organizations, NGOs, etc.*: Inviting apex bodies of government and private organizations, NGOs, etc. in seminars, symposia, workshops which will be helpful in popularizing Home Science as a competitive stream. More over the misconcepts

about Home Science can be cleared only when discussions will be made at such platforms.

4. *Integrated research of students at renowned institutions*: Integrating researches of various departments of Home Science with renowned institutes will be helpful to get recognition of home science researches in other related fields. Nutrition department can integrate research with organizations like NIN, ICMR; WFP etc. Similarly researches of Human Development and Extension can be integrated with NGOs like UNICEF, CARE, CRY, CHETNA and department of Human Resource Management can combine research with Consumer forum at district and state level and similarly researches of department of Clothing and Textile can be integrated with Textile industries. This will help people to know the type of researches being done at the college.

5. *Initializing programmes on TV and Radio that includes Home Scientists*: Programmes should be initialized on TV and Radio that include Home Scientists to popularize work of home science. Cookery programme can be a good example of disseminating scientific knowledge along with skill. As Home Scientist will not only demonstrate to cook tasty food but they explain the nutrition value and calories associated with it. Similarly programmes like healthy living can be started in which experts from various departments of Home Science can be called upon to give suggestions on various aspects of home and living such as home decoration, healthy food, garment designing, child care, teen

age adjustments etc. This will help to popularizing various departmental activities.

6. *Advocating about professional and entrepreneurial opportunities associated with Home Science studies* (as discussed above)
7. *Change in the nomenclature and thus distinguishing Home Science stream from arts stream*: The name of the courses of Arts stream and Home Science is similar that creates confusion in the mind of the people. So change in nomenclature will help in distinguishing Home Science stream from arts stream.

Home Science Extension and Rural Development

Chapter 3

Introduction to Home Science Extension

Extension

The word extension is derived from Latin word *'ex'* and *'tensio'* that means 'out stretching'. Extension can be defined as "science of developing capability of people for sustainable improvement in their quality of life". (Sustainable means management of resources to satisfy changing human needs, while, maintaining or enhancing quality of environment and conserving natural resources.)

The word extension, in true sense was first used in a practical way by James Stuart, also known as "Father of University Extension; of Trinity College, Cambridge University in 1867-68, while addressing "Women Association and Working Men's Clubs" of North England. Dr. K.N Singh is Father of Extension Education (India).

Extension Education

An Extension Education is a dynamic; need based, problem oriented, non-formal system for budding and implementing programs for development of the people. It is an applied behavioural science, the knowledge of which is applied to bring about desirable changes in the behavioural complex of human beings usually by applying the latest scientific and technological innovations.

Extension education educates rural people outside the regularly organized school and class rooms for bringing out social and cultural development. It disseminates information to people who needs it.

More Definitions

B. Rambhai (1958)

Extension is a two-way channel; that brings scientific information to the village people, and also takes the problems of the village people to the scientific institution for solution. It is a continuous educational process, in which both learner and teacher contribute and receive.

J.P. Leagans (1961)

Extension education is an applied science consisting of content derived from research, accumulated field experiences and relevant principles drawn from the behavioural science synthesised with useful technology into a body of philosophy, principles, content and methods focussed on the problems of out of school education for adults and youth.

Extension education is the process of teaching rural people how to live better by learning ways that improve their farm, home and community institutions.

D. Ensminger (1961)

Extension is a programme and a process of helping village people to help themselves, increase their production and to raise their general standard of living.

Kelsey and Harne (1963)

Extension work is an out of school system of education in which adult and young people learn by doing. It is partnership between the Govt. and the people, which provides service and education designed to meet the people. Its fundamental objective is the development of the people.

O.P. Dahama (1973)

Extension Education is defined as an educational process to provide knowledge to the rural people about the improved practices in a convincing manner and help them to take decision within their specific local conditions.

Thus, it can be said that extension is means to bring out inevitable change in the right direction because change is necessary to make the world better place to live in.

Formal, Non-formal and Informal Education

Formal Education is organized, highly institutionalized, chronically graded, and hierarchically structured with fixed curriculum. Here the knowledge flows from teacher to learners and learning starts from theoretical aspect which is then applied to practical. Example: Education at School and College.

Non-formal Education is organized systematic educational activity carried outside framework of formal system. It provides selected type of learning to particular group of population. Example: Adult Education

Difference between Formal and Non-formal Education

Sl.No.	*Formal Education*	*Non-formal Education*
1.	Teaching in premises of institution	Teaching is outside four wall of institution
2.	Learners are homogenous	Learners are heterogeneous
3.	Learners have common goal	Learners have diverse goals
4.	There is no free choice for learners. Institutional norms are followed	Learners have freedom to chose subject for learning
5.	There is fixed curriculum	No fixed curriculum
6.	It is more theoretical	It is more practical and intended for immediate application in problem solving.
7.	Starts from theoretical aspect and then applied in practicals	It is practo-theoretical that means practical leads to theoretical concept
8.	It brings change in the knowledge level of students	It brings change in the human behaviour
9.	Knowledge flows from teacher to learners	Knowledge flows from teacher to learners and learner to teacher.
10.	Teaching is mainly vertical	Teaching is mainly horizontal
11.	Class attendance is compulsory	Participation is voluntary.
12.	Formal education is rigid	Non-formal Education is flexible
13.	Example: School or College Education	Example: Extension Education

Informal Education is life long process by which every person acquires and accumulates knowledge, skill and attitude from daily experience and at home from mother, at office or from friends in play group *etc.*

Home Science Extension

Home Science Extension is type of education through which knowledge and skill, related to different areas of Home Science, is extended to those who are deprived of formal education, *i.e.* scientific and technological knowledge.

Home Science extension is multi-disciplinary in nature since it involves various aspect of development. It is applied science which aims at bringing change in behavior of target group through dissemination of scientific and technical information.

Philosophy of Home Science Extension

The word philosophy is derived from two Greek words 'Philos' and 'Sophia'. 'Philos' means 'knowledge' and 'Sophia' means 'manner'. In this way the correct meaning of philosophy is the manner of achieving the knowledge or comprehensive idea or point of view about anything. The Philosophy of Home Science Extension is "Helping people to help themselves become superior human being". Philosophy of Home Science extension education encourages a person to bring about his own development and that of society through his own leadership and motivation by following scientific approach and democratic ways.

Philosophy of Extension According to Einsminger (1961)

- ☆ Extension is an educational process. Extension wants to change the knowledge, attitudes, understanding and skill of the people.
- ☆ Extension is working with men, women, youths, boys and girls to answer their needs and their wants. Extension is teaching people what to want and ways to satisfy their wants.
- ☆ Extension is "helping people to help themselves."
- ☆ Extension is working on the basis of learning by doing and seeing.
- ☆ Extension wants development of individuals, their leaders, their society and their world as a whole.
- ☆ Extension is working together to expand the welfare and happiness of the people.
- ☆ Extension is working in harmony with the culture of the people.
- ☆ Extension is a two-way channel process.
- ☆ Extension is a continuous educational process.

According to Dharma, Philosophy of Extension is Based on

- ☆ Self help
- ☆ People as resource
- ☆ Cooperative effort in problem solving
- ☆ Foundation for democracy
- ☆ Having two ways channel
- ☆ Creating need and interest by seeing and doing
- ☆ Is voluntary in nature

- ☆ Ends are achieved through persuasion and education
- ☆ Programme is based on attitude and value of people
- ☆ Is never ending process

Characteristics of Home Science Extension

1. It uses a multidisciplinary approach: As it draws knowledge from various disciplines such as textile, nutrition, human resource management, human development, extension and communication.
2. It is action oriented: It means Home Science Extension conduct series of activities such as discussion, lecture, demonstrations etc. to impart knowledge to home maker.
3. Mainly women are its target group: Though adolescent girls, boys, children are included as participant in extension activities but women are focused group of home scientist as women are key persons in family and also because, it is well said that, when we teach a female we teach a family and not an individual.
4. Its results are intangible: Main aim of Home Science Extension is to bring behaviour change (which are intangible in nature) *i.e.,* it cannot be seen.
5. It is two way process: *I*t imparts scientific and technical knowledge from the University to the target group. And also takes feedback from target group to scientists at the University, for further improvement in technology.

Principle of Home Science Extension

Principle is statement of policy to guide decision and action. There are eleven principles of Home Science Extension.

1. Principle of culture difference
2. Grass root principle
3. Principle of indigenous knowledge
4. Principle of need and interest
5. Principle of learning by doing
6. Principle of participation
7. Family principle
8. Principle of leadership
9. Principle of adaptability
10. Principle of satisfaction
11. Principle of evaluation

1. Principle of Culture Difference

Culture difference means there is variation in customs, values, attitude and way of living of the people. Extension work to be successful must carry out in harmony with culture pattern of the people because rural people have high faith in their traditional customs and values.

2. Grass Root Principle

Grass root means what people have and where they are. Extension programme should start with local situation, local groups, and local problems. Change should start from existing situation and with resources available with them.

3. Principle of Indigenous Knowledge

Indigenous means traditional knowledge which people

have developed through generations of work experience. Extension agent should try to understand their indigenous/traditional knowledge before recommending them something new.

4. Principle of Need and Interest

For making extension successful the extension agent should identify need of the people and plan the programme accordingly. This will generate interest amongst the people for the programme and will increase their cooperation in it. It is also important to know that there cannot be one programme for all people as the needs and interests of people differ from individual to individual, from village to village, from block to block, and from state to state.

5. Principle of Learning by Doing

Doing work by own-self helps people to gain confidence and learn more because it involves maximum sensory organs. So, extension should give opportunity to rural people to practically apply the new information or technology.

6. Principle of Participation

Good extension work is directed towards assisting rural families to work out their own problems rather than giving them ready-made solutions. Extension worker should involve rural people in extension programme right from the planning stage so that they should feel that it is their programme. Actual participation and experience of people in these programmes creates self-confidence in them and also they learn more by doing.

7. Family Principle

Extension work have a better chance of success if the extension workers have a whole-family approach instead of separate and unintegrated approach. Thus target of extension agent should be development of family as a whole *i.e.,* women, youth, children, men *etc.*

8. Principle of Leadership

Role of extension agent should be to develop leadership among rural people and to identify different types of leader among them. This selection and training of local leaders to enable them to help to carry out extension work is essential to the success of the programme. People have more faith in local leaders and they should be used to put across a new idea so that it is accepted with the least resistance.

9. Principle of Adaptability

Extension teaching method should be flexible and adapted to suit local condition because people, their situation, resources varied from place to place. Flexibility in programme helps to bring necessary changes whenever needed, to meet the varying conditions.

10. Principle of Satisfaction

At the end of extension programme, people should get satisfaction as a result of solving problem, meeting a need, acquiring a new skill or some other changes in behaviour because this satisfaction results reinforce learning and motivate people to seek further improvement. In extension work it is said "A satisfied customer is the best advertisement." So satisfaction is the key to success in extension.

11. Principle of Evaluation

Extension programme should be evaluated. The effectiveness of the work is measured in terms of the changes brought about in the knowledge, skill, and attitude and adoption behaviour of the people but not merely in terms of achievement of physical targets. Evaluation should indicate gaps and steps to be taken for further improvement. Evaluation prevents stagnation.

Need for Home Science Extension

It was seen that women were not participating in development so need for Home Science Extension emerged. It was felt that women can only contribute in development of nation when the technological gap, of what women have or what she knows will be filled with what they should have or what they should know. And this gap could be filled with the help of extension activities of extension workers, as extension worker is link between the researcher and the local people.

It is essential for extension agent to know the various techniques to teach so that they can encourage people to adopt new research techniques easily.

Merely awareness about objectives of extension method by home scientist does not solve any problem; rather a good extension scientist should be aware of local problems, local situations, local resources and feasibility. After thorough analysis of the situation, information about innovative scientific techniques should be given to the people for adoption. Therefore, the need for studying Home Science Extension becomes important, so that they can encourage the adoption of new techniques to reduce loss of resources at home level and increasing efficiency and efficacy.

Objective of Home Science Extension

The objectives are the expressions of the ends towards which our efforts are directed or in other words objective is direction of movement. The fundamental objective of extension education is the development of the people. Home Science extension in our country is primarily concerned with following main objectives:

- ☆ To assist women to discover their problems and identify felt need.
- ☆ To make women aware of women's rights
- ☆ To make her aware of consumer right
- ☆ To assist women in mobilizing and utilizing the resources they have.
- ☆ To raise the standard of living of women
- ☆ To develop the feeling of self-dependence among rural women.
- ☆ To increase the net income of farm women by capacity building
- ☆ To assist women to develop leadership among them.
- ☆ To disseminate research information home maker.
- ☆ To collect feedback information from rural to the research institution.
- ☆ To increase the facilities for social, cultural and entertainment programmes for rural women
- ☆ To improve educational level of women
- ☆ To improve social condition of women
- ☆ To encourage rural people to participate in community programmes

Chapter 4
Role of Home Science in Rural Development

Rural Development

Rural development is dynamic process that aims in complete transformation of rural life for betterment. It is mainly concerned with progress or growth of rural people. Rural Development includes agriculture development, change in social, economic and political infrastructure, women and child development, public health, education, nutrition *etc.* It emphasizes on 'helping people to help themselves'. Its effort is directed towards creating awareness among rural people and uniting them to solve their problems collectively.

Constraints in Rural Development

- ☆ Rural poverty leads to low input capacity of people thus the output is less.

- ☆ Fragmented land holding hinders in inclusion of new technology in the farm thus agriculture development lag behind
- ☆ Efforts for rural development by government and private sectors are like drop in ocean, since the population is high.
- ☆ Illiteracy leads to difficulty in making them understand the innovations thus reduces the rate of adoption of innovations
- ☆ Most of the programmes for family and community development are fragmented and limited in scale
- ☆ Lack of coordination between various sectors (government/private) and different programmes that work for rural development.
- ☆ Rural development programmes are mostly poor quality and programme implementation is haphazard
- ☆ Rural development programmes mainly concentrate on large farmers so small farmers are neglected
- ☆ Rural development can take place only when development starts from home and family but majority of rural development programmes concentrates only on agriculture development.
- ☆ Women who contribute in various chores of farm, household and animal husbandry are deprived of training and other benefits from government/ private sector.

Importance of Home Science Extension in Rural Development

Home Science Extension is primarily for women and child development. Its main objective is to empower women and increase their status in society. Extension education is an educational process by which capabilities of people, especially women, are developed to understand their own needs, problems and find solution to them. Extension helps to avail scientific methods to the rural women, so that they can raise their standard of living by means of improving nutritional status, maintaining health and hygiene, making them aware of their rights and helping them to solve their problems scientifically within their own sphere of resources. Community development is possible only by bringing change in the behaviour complex of rural women and extension education plays major role in bringing desirable change. Through extension activities knowledge and skill of researches done at university level is transferred to the farm women.

Chapter 5

Historical Developments in Extension Education

The development of extension in India has seen a large number of experiments. For understanding the development of systems of extension in India it is important to have knowledge of early extension efforts.

Pre-Independence Efforts in Development of Extension

Pre-independent efforts were attempted by some benevolent persons and private agencies to improve rural life. Few programmes are listed below:

Sl.No.	*Project/Programme*	*Year*	*Initiated by*
1.	Shriniketan	(1921)	Rabindra Nath Tagore
2.	Martandum	(1921)	Dr. Spencer Hatch
3.	Gurgaon Project	(1920)	Bryne, F.L.
4.	Sevagram or Gandhian Constructive Programme	(1920)	Under Guidance of Mahatma Gandhi

1. Shriniketan (1921)

This project was started by Rabindra Nath Tagore in collaboration with Elmhurst in Bengal in the year 1921. The aim of project was to:

- ☆ Win friendship of villagers and to make honest effort to assist them in solving their problems
- ☆ Develop people's resources and their capacity
- ☆ Teach them better methods of growing crops, vegetables and keeping livestock
- ☆ Encourage villagers to learn and practice art and craft

The project did not succeeded because of lack of fund to finance the expanded activities and also because project was idealistic and practical aspect of project was neglected.

2. Martandum (1921)

- ☆ This project was started by Dr. Spencer Hatch in Ravenscourt State under auspices of Y.M.C.A. in the year 1921.
- ☆ Its objective was to bring complete upward development towards a more abundant life for rural people in spiritual, mental, physical, social and economic fields.
- ☆ It gave expert guidance, training to person to improve their performance, making people conscious of their wants and needs.
- ☆ The inadequate fund and Government help and religious bias lead to failure of the project.

3. Gurgaon Project (1920)

- It was organized Bryne, F.L. (He was Collector of Gurgaon District in Punjab).
- The objective of the project was to emphasize on increasing farm yield, cutting expenditure on social and religious functions, improving health standards, organizing welfare programme.
- The major shortcoming of the project was that it could not develop leadership in village who would continue work when the village guides had left the villages

4. Sevagram or Gandhian Constructive Programme (1920)

- It was started under guidance of Mahatma Gandhi.
- Its objectives were to provide service to underprivileged, achieve self dependency and provide basic education to the people.
- Activities of project were organization of training centers for cottage industries, prohibition of untouchability, stressing women education.
- Very high personal standards were set in projects which were difficult to achieve. So, project could not provide expected result.

Shortcomings of Pre-Independence Extension Projects

- Efforts were based on individual initiative
- Lack of support from the government
- All attempts were unseen and discontinuous

- Inexperienced and untrained staff
- Unbalanced and ill defined programmes
- Need for proper method and skill of approach was not realized
- Evaluation was not carried out, hence results are not known
- Lack of association and coordination of other development departments
- Local people were not involved in planning and execution of the programme.

Post-Independence Effort in Development of Extension

After independence there was realization of need for social transformation. So projects related to social transformation were mainly emphasized.

1. Firka Development Scheme (1946)

It was started in Madras and it aimed at attainment of Gandhian ideal of "Village Swaraj". Its aim was not only educational needs but also to improve economic, sanitation to make people self-confident and self-reliant. It was later merged with National Extension Service in 1953

2. Etawah Pilot Project (1948)

- It was organized by American architect Mr. Albert Mayer in 1948.
- Its objective was to see extent of improvement possible in production, social improvement, self-reliance and co-operation.
- It was also intended to see whether results obtained could be permanent and introducible elsewhere.

- The activities were organized with help of trained village level workers and co-operation of other departments and agencies were enlisted. The Demonstration method was used to convey new ideas to the people.
- The project proved to be successful and on a similar pattern Extension Organization was planned in India.

3. Nilokheri Experiment (1948)

- Nilokheri Experiment was done by Shri S.K. Dey a foreign trained engineer at Kurukshetra.
- He worked for rehabilitation of people during partition by giving them work, so that they could earn. For this he made survey of camp population and enlisted their requirements. Clothing was found to be the first requirement thus a scheme was prepared to start a small training-cum-production centre.
- Its immediate aim was to teach them weaving, tailoring, printing *etc.* and later started small production centre.
- It set example for the others to follow.

4. Comilla Project (1987)

Comilla Project was carried on at Kotwali Thana in Bangladesh. Its objectives were:

- To improve rural living standards by increasing productivity of land by more intensive cultivation
- To improve the roads, irrigation and drainage system

- ✰ To set up an organization to introduce improvement to farmers and provide training

It showed how two-way communication between farmer and outside source is effective then one-way or top-down approach of the Government. This project aroused interest of rural development workers all over the world.

Chapter 6
Community Development Programme and Panchayati Raj Institutions

Community Development Programme

The term Community Development appears to have originated from Cambridge in England in 1948. Community development is process of change from traditional way of living of rural communities to progressive way of living. It focuses on development of people. It is method by which people can be assisted to develop themselves on their own capacity and resources. It is dynamic, unified problem solving approach. Community Development Programme was launched in India on 2nd October, 1952. It started with 15 pilot projects all over India. By 1963 Project spread to whole country. Name of Community Development was later changed to Rural Development Programme.

Objectives of Community Development Programme

- Developmental of community as a whole, with special emphasis on weaker and underprivileged sections
- Area Development with minimum all round progress in every field.
- Seeking people's participation in development process
- To assist each village to have effective Panchayat, Cooperatives, Schools *etc.*
- Increase agriculture production
- Improve existing village craft, industries *etc.*
- To provide minimum, essential health services and improving health practices
- Providing educational facilities for children
- Providing recreational facilities
- Improving housing and family living conditions
- Providing programme for village women and youth

Reason of Failure of Community Development Programme

It was carried out at large number of places, so it became difficult to recruit and train enough people to play all essential roles. It was not feasible. A non-political approach for transformation resulted in helping only those who were powerful in village led to failure of the Community Development Programme.

Panchayati Raj Institution

Panchayat is ancient Indian word that means five people or Headman. Thus it can be said that Panchayat is council of five elders. On recommendation of Balwant Rai Mehta Committee "Panchayati Raj System" was introduced in India in 1958. Panchayat first started in Nagaur, Rajasthan and Andhra Pradesh in 1959.

With the passage of Constitution (7311 Amendment) Act, in 1992 an Act came into force on 24 April, 1993 that stipulated: Panchayats shall be given powers and authority to function as institutions of self-government.

Now Panchayati Raj system exists in all the states except Nagaland, Meghalaya and Mizoram.

Salient Features of the Act

- ☆ A three-tier system of Panchayati Raj for all states having a population of over 20 lakhs.
- ☆ Regular Panchayat elections every 5 years
- ☆ Reservation of seats for Scheduled Castes, Scheduled Tribes in proportion to their population and not less than 1/3 seats for women
- ☆ State Finance Commission to make recommendations on financial powers of the Panchayats
- ☆ Independent State Election Commissions

Powers and Responsibilities of Panchayats

- ☆ Preparation of plan for economic development and social justice.
- ☆ Implementation of schemes for economic development and social justice (in relation to 29

subjects given in Eleventh Schedule of the Constitution).

- ☆ To levy, collect an appropriate taxes, duties, tolls and fees.

Structure of Panchayati Raj Institution

Panchayati Raj System is three tier system *i.e.*, it has three levels: Village Panchayat, Block Panchayat and District Panchayat. Panchayat Raj System of governance in which Gram Panchayat is basic unit of administration.

Village Panchayat

At village level it is called Panchayat. Village Panchayat is local body working for well being of the village. The number of members ranges from 7 – 31. Village Panchayat is headed by "Sarpanch" who is elected by members of village Panchayat. The functions of village Panchayat are: public health and sanitation, water supply, street light, maternity and child welfare, registration of birth and death, education, recreation, construction *etc.*

Intermediate Level Panchayat

The block level institution is called "Panchayat Samiti". It is also called "Block Panchayat". It is Panchayat set at block/tehsil in state where total population exceeds 20 lakh.

The members of Panchayat samiti are counselor of zilla parishad elected from electoral division included in the block. They are entrusted with development function and are directly responsible for implementation of community development programme. They are also charged with preparation and implementation of developmental plans for the block.

The samiti is vested with (given) specific executive responsibilities in matter of primary education, health, sanitation and communication. They also supervise the work of village Panchayat and have right to scrutinize the budget of Panchayats.

Zilla Panchayat

Zilla Panchayat is also called district level Panchayat. It is headed by President. The members of zilla parishad consist of representatives of Panchayat samitis, members directly chosen by people *etc.*

The zilla parishad exercise general supervision over the working of Panchayat samities and advised government on implementation of the development scheme. Beside these duties, it has specific executive functions in establishment maintenance and expansion of secondary, vocational and industrial school.

Chapter 7

Women and Child Development Programmes

Women population in our country is almost half of our total population and their count are more than the total population of many other countries. Many policies and programmes are run by government and non-government agencies for upliftment of women. To help women avail these opportunities the extension agent need to be aware of these programmes and extend its information to the rural women so that they can gain maximum benefit out of it.

Women and Child Development Programmes

- ☆ Applied Nutrition Programme (ANP): 1959.
- ☆ Integrated Child Development Scheme (ICDS) : 1975
- ☆ Training of Rural Youth for Self-Employment (TRYSEM): 1979

- ☆ Development of Women and Children in Rural Areas (DWACRA) : 1984
- ☆ Balika Samruddhi Yojana : 1997
- ☆ Suvarn Jayanti Gram Swarojgar Yojana : (SJGSY) 1999-2000
- ☆ RMK (Rashtriya Mahila Kosh) : 1993
- ☆ Indira Mahila Yojana (IMY): 1995
- ☆ Short Stay Homes (SSH)
- ☆ Sampoorna Grameen Rozgar Yojana: 2001

1. Applied Nutrition Programme (ANP): 1959

It is sponsored by United National Children's Fund and Central Government of India. Its goal is directed towards improving health of children and expectant mothers by supplementary diets (Protein, Vitamin, Rich food, Milk, Vegetable, and Egg *etc.*). UNICEF supports the Applied Nutrition Programme by providing equipment and supplies and trainees.

The programmes started under ANP are:

- ☆ Expanded aid to nutrition programme
- ☆ Vitamin "A" prophylaxis programme was started to prevent blindness in children
- ☆ Anemia prophylaxis programme
- ☆ Special/supplementary Nutrition programme for infant, toddlers and pre-schooler

2. Integrated Child Development Scheme (ICDS): 1975

ICDS Scheme was launched on 2nd October 1975. It represents one of the world's largest and most unique programmes for early childhood development. It is Central Government financed programme that provides pre-school education and also breaks the vicious cycle of malnutrition, morbidity, reduced learning capacity and mortality among children.

The Integrated Child Development Services (ICDS) Scheme was launched in 1975 with the following objectives:

1. To improve the nutritional and health status of children in the age-group 0-6 years;
2. To lay the foundation for proper psychological, physical and social development of the child;
3. To reduce the incidence of mortality, morbidity, malnutrition and school dropout;
4. To achieve effective co-ordination of policy and implementation amongst the various departments to promote child development; and
5. To enhance the capability of the mother to look after the normal health and nutritional needs of the child through proper nutrition and health education.

The above objectives are sought to be achieved through a package of services comprising of:

1. Supplementary nutrition,
2. Immunization,
3. Health check-up,
4. Referral services,
5. Pre-school non-formal education and
6. Nutrition and health education.

The Beneficiaries of ICDS programme are:

- ☆ 0-6 years of children
- ☆ Pregnant and lactating mother
- ☆ Community women 15-45 years of age

3. Training of Rural Youth for Self-Employment [TRYSEM]: 1979

It was launched in August 15th, 1979. The main aim of this programme is to equip rural youth with necessary skill and techniques to enable them to take vocations or self-employment. The beneficiary are- Rural youth of 18-35 age groups. Provision was made that amongst total rural youth 30 per cent from Scheduled Cast and Scheduled Tribe (SC and ST) categories and minimum one-third should be women.

4. Development of Women and Children in Rural Areas (DWACRA): 1984

The programme started in 1984 and is running under IRDP (Integrated Rural Development Programme). The main aim of this programme is to improve status of under privileged women. It provided income generating activities to women. It also promotes training of women in income generating activities, a concurrent market for their product, easy loan accessibility to women and creating support system through crèches and day-care-centre

5. RMK (Rashtriya Mahila Kosh) : 1993

The National Credit Fund for Women or the Rashtriya Mahila Kosh (RMK) was set up in March 1993 as an independent registered society by the Department of Women and Child Development in Government of India's

Ministry of Human Resource Development. It is the most prominent national level micro-finance apex organization providing micro-finance services for women in India. The National Credit Fund for Women is an innovative mechanism for reaching credit to poor women. Through access to credit, it aims to raise the capacity of women by enhancing through productivity and economic self- reliance. It has provided credits to over 2.32 lakh women since its inception from 1993. It encourages formation of Self Help Groups (SHGs) for promotion of thrift and credit leading to income generation activities.

Its main objectives are:

- ☆ To provide or promote the provision of micro-credit to poor women for income generation activities or for asset creation.
- ☆ To adopt a quasi-informal delivery system, which is client friendly, uses simple and minimal procedures, disburses quickly and repeatedly, has flexibility of approach, links thrift and savings with credit and has low transaction costs both for the borrower and for the lender.
- ☆ To demonstrate and replicate participatory approaches in the organization of women's groups for thrift and savings and effective utilization of credit.
- ☆ To use the group concept and the provision of credit as an instrument of women's empowerment, socio-economic change and development.
- ☆ To cooperate with and secure the cooperation of the Government of India, State Governments, Union Territory administrations, credit institutions,

industrial and commercial organizations, NGOs and others in promoting the objectives of the Kosh.

- ✰ To disseminate information and experience among all these above agencies in the Government and non-government sectors in the area of microfinance for poor women.
- ✰ To receive grants, donations, loans, etc., for the furtherance of the aims and objectives of the Kosh.

6. Indira Mahila Yojana (IMY): 1995

The scheme is related with women's coordinated development. The scheme is sponsored by the Central Government and implemented by the State Government.

Main aim is to motivate capabilities of women especially earning capacity, awareness of their demands *i.e.,* to impart women in Panchayat, to improve Panchayat tradition in favor of women. It is essential to establish (IMY) at all Anganwadis by which women can unite and develop through elder's education and health care programme.

7. Balika Samruddhi Yojana: 1997

The scheme initiated in 1995 modified in 1997. The beneficiaries of scheme is female children. Its objective is to change attribute of common society towards the female child. Earlier it gave Rs.500 for girl's birth. Now Rs.500 is given as scholarship after admission of girl child to school. Its aim is to enhance percentage of Balika education.

8. Suwarnjayanthi Gram Swarozgar Yojana (SJGSY): 1999-2000

Launched on 1st April, 1999, the programme replaces the previous self-employment and allied programmes -

IRDP, TRYSEM, DWCRA, SITRA, GKY and MWS, which are no longer in operation. SGSY is a centrally sponsored scheme and funding is shared by the Central and State Governments in the ratio of 75:25.

SGSY is a credit-cum-subsidy programme, wherein credit is the key component. It covers all aspects of self-employment, such as organization of the poor into Self-Help Groups (SHGs), training, credit, technology, infrastructure and marketing.

The objective of Swarnjayanti Gram Swarozgar Yojana (SGSY) is to bring the assisted poor families (Swarozgaris) above the poverty line by organising them into Self-Help Groups (SHGs) through the process of social mobilisation, their training and capacity building and provision of income generating assets through a mix of Bank credit and Government subsidy.

SGSY lays emphasis on activity clusters. About 10 activities are identified for each block with the approval of Panchayat Samitis, however emphasis are on 4-5 activities for cluster development.

The programme covers families below the poverty line in rural areas of the country. Within the target groups special safeguards have been provided for the benefit of the vulnerable groups *i.e.* SCs/STs, women and physically handicapped persons. Accordingly, SCs/STs should account for 50 per cent of the Swarozgaris, women for 40 per cent and the disabled persons for 3 per cent.

The Gram Sabha will authenticate the list of families below the poverty line identified in BPL census. Identification of individual families suitable for each key activity will be made through a participatory process involving Gram Panchayat, Block functionaries (*i.e.* BDO or his representatives) and the banker.

9. Vande Mataram Yojana: 2004

It is unique partnership between the Government of India and Private Sector. It was started on 23rd February, 2004. Aim of scheme is to improve health status of women and her children during the period of pregnancy. In this scheme Doctors of Federation of Obstetricians and Gynecologist Society of India (FOGSI) gather on 9th day of every month for free check-up of expectant mother. It also included ante-natal and post-natal cure, free iron tablets contraceptives were also available

10. Short Stay Homes (S.S.H.)

SSH protects and rehabilitate those women and girls who are facing social, economic and emotional problems due to family problems, mental stress, social ostracism, exploitation or other causes or are being forced into prostitution and are in moral danger. The scheme provides following services/facilities.

- Temporary shelter to the needy women and girls from six months to three years
- Case work and counseling services
- Medical care and psychiatric treatment
- Occupational therapy, skill development training and rehabilitation services
- Educational, vocational and recreational activities
- Follow up by counselors for those women who have left the Homes, needs to be done to ensure their well-being and safety

Children accompanying the mother or born in the Home, may be permitted to stay in the Home up to the age of seven years after which, they may be transferred to children's institutions or provided foster care services.

11. Sampoorna Grameen Rozgar Yojana: 2001

It is Centrally Sponsored Scheme, namely, Sampoorna Grameen Rozgar Yojana (SGRY). It was launched on 25th September, 2001 by merging the on-going schemes of EAS and the JGSY.

The objectives of the programme is to provide additional wage employment in the rural areas as also food security alongside the creation of durable community, social and economic infrastructure in the rural areas. The programme is self-targeting in nature with special emphasis to provide wage employment to women, scheduled castes, scheduled tribes and parents of children withdrawn from hazardous occupations.

The Programme is being implemented in two streams:

1. The First Stream of the Programme will be implemented at the District and Intermediate level. Panchayats: Fifty per cent of the funds available under the SGRY are earmarked for First Stream; these are distributed between the Zilla Parisad and the Intermediate Panchayats in the ratio of 40:60.
2. The Second Stream of the Programme will be implemented at the Village Panchayat level. Fifty per cent of the SGRY funds are earmarked for this Stream. The entire funds are released to the Village Panchayats through the DRDAs/Zilia Parishads.

The salient features of the proposed Scheme are as under:

- ☆ The Sampoorna Grameen Rozgar Yojana (SGRY) is a Centrally Sponsored Scheme (CSS) being implemented with a total outlay of Rs.10,000 crores.
- ☆ Under the Scheme, 50 lakh tonnes of food grains amounting to Rs.5, 000 crores (at economic cost) will be provided every year, free of cost to the State Governments and Union Territory Administrations.
- ☆ The remaining funds (Rs.5, 000 crores) will be utilized, to meet the cost component of wages and material cost.
- ☆ The cost of the cash component of the Programme will be shared by the Centre and States in the ratio of 75:25.
- ☆ The payment for food grains will be made by the Ministry of Rural Development to the Food Corporation of India (FCI) directly.
- ☆ About 100 crores mandals of employment are envisaged to be generated every year in the rural areas through the SGRY.
- ☆ Fifty per cent of the total available funds under the SGRY will be provided to each stream.
- ☆ Every worker seeking employment under the SGRY will be provided 5 kg. of food grains (in kind) per mandal as part of wages.
- ☆ The balance of wages will be paid in cash so that they are assured of the notified minimum wages.

- ✰ The State Governments and UT Administrations will be free to calculate the cost of food grains (paid as part of wages) at either BPL rates or APL rates or anywhere between these two rates.
- ✰ The SGRY has been in operation from the financial year 2002-2003. Since the Scheme was launched in the middle of 2001-02, the ongoing schemes of the EAS and the JGSY were merged with the new Scheme.

The Programme permits works which lead to the creation of additional wage employment, durable assets and infrastructure, particularly those which assist in drought proofing such as soil and moisture conservation works, watershed development, promotion of traditional water resources, afforestation and construction of rural infrastructure and link roads, primary school buildings, dispensaries, veterinary hospitals, marketing infrastructure and Panchayat Ghars in rural areas.

The overall supervision of the programme rests with the Zilla Parisad. The programme is be regularly monitored by the Department of Rural Development in accordance with the in-built monitoring mechanisms, including periodical reports and returns, Vigilance and Monitoring Committees, visits by officers of the Central and State Governments and by the Area officers of the Ministry of Rural Development. The programme is also evaluated through the studies conducted by reputed institutions, organizations and sponsored by the Central/State Governments.

Chapter 8
Agriculture Development Programmes for Rural Families

In India more than 75 per cent of population live in villages, thus for India to develop it needs to develop rural areas first. An extension agent should know the various schemes available for the poor to improve their socio-economic status. There are various programmes for development of agriculture such as:

Agriculture Development Programme

- IADP: Intensive Agriculture District Programme (1960)
- HYVP: High Yielding Varieties Programme (1966)
- IVLP: Institution Village Linkage Programme (1999)

- ☆ NATP: National Agriculture Technology Project (1990)
- ☆ ATMA: Agricultural Technology Management Agency (2000)
- ☆ ATIC: Agriculture Technology Information Centre (2000)
- ☆ NAIP: National Agricultural Innovation Project (2000)

1. Intensive Agriculture District Programme (IADP): 1960

The steep fall in crop productions in the drought year 1957-58 focused attention on the seriousness of the food situation up substantially. In these circumstances, the government of India invited a team of agriculture experts sponsored by the Ford Foundation to make a careful study of Indian agriculture and make recommendations future actions. The Government of India accepted in general, the recommendations made by this team and IADP was started. It was also called 'Package programme'

The programme started in 1960 in Seven Districts namely: Thanjavur (Tamil Nadu), West Godavari (Andhra Pradesh), Shahabad (Bihar), Raipur (Madhya Pradesh), Aligarh (Uttar Pradesh), Ludhiana (Punjab) and Pali (Rajasthan).

The major objectives of the programme were to demonstrate the most effective ways of achieving rapid and significant increase in agriculture production through adoption of a package of improved practices and to suggest successful and adoptable innovations, approaches and procedures which could be adopted in similar areas of the country.

Its impact was less because production efforts were widely scattered over entire country.

2. High Yielding Varieties Programmes (HYVP): 1966

IADP concerned on promotion of intensive agriculture, they operated within the limitations set by the existing crop varieties which had low fertilizer response.

In the year 1963 - 64 a few Mexican dwarf wheat varieties were tried out on selected basis and in 1964 - 65 an exotic variety of paddy, namely, Taichung native - 1 was tried and the results were quite encouraging. Hybrid maize, hybrid bajra and several varieties of improved millets were also in advance stage of development.

In 1966 a new programme was instituted cultivation of high yielding varieties which was known as the high - yielding variety programme. The programme was to be taken up in a phased manner by the state government.

Its objectives was to boost up agricultural production by using high yielding variety with appropriate inputs.

It was a strategy under Fourth Five Year Plan which aimed at attaining self sufficiency in cereal foods by end of 1970-71 by introducing High Yielding Varieties in cereals.

The HYVP was the responsibility of the state government for its implementation from its existing resources under the guidance and supervision of the central government. No separate funds were allocated for this at any level. The area under high yielding varieties increased from six million hectares in 1967 - 68 to 34 million hectares in 1967 - 77.

3. National Agriculture Technology Project (NATP): 1990

NATP was initiated to consolidate earlier investment. The basic premise of NATP is that research and extension programmes should be farmer centered and demand driven.

Broadly speaking NATP has three major components, *viz.,* Development of ICAR (Indian Council of Agricultural Research) organization and management system; Support for agro-eco-systems research; Innovations in Technology Dissemination.

4. Institution Village Linkage Programme (IVLP): (1999)

It was seen that once the technology was developed, laboratory/research stations thought their job to be over. There was little linkage of this research station with actual adoption of these technologies on the farm of the farmers. For this purpose there arouse need to conduct "On Farm Research". To fulfill the gaps in linkage between Research – Extension – Farmers and to overcome lacuna in TOT (Transfer of Technology); TAR – ILVP under NATP was launched in 1999. The methodologies adopted in this programme is PRA (Participatory Rural Appraisal)

Its objectives were:

- ☆ To assess the existing recommended technologies through Verification Trials (V.T.) on the farmer's field.
- ☆ To refine the recommended technologies by conducting 'On-Farm-Trials' (OFT) after assessment.

- ☆ To arrange suitable need based trainings of farmers, youth and farm women in order to develop skill and confidence and provide self-employment.

5. Agricultural Technology Management Agency (ATMA): 2000

ATMA is a society of key stake holders involved in agricultural activities for sustainable agricultural development in the districts.

Its objectives are:

- ☆ To decentralize decision making to district level through the creation of Agricultural Technology Management Agency (ATMA)
- ☆ To increase farmer input into programme planning and resources allocation
- ☆ To increase programme coordination and integration so that programme thrust such as farming system innovative farmer organization technology gap can be more effectively and efficiently be implemented.

6. Agriculture Technology Information Centre (ATIC): 2000

The farmers are often not aware as to whom and where to approach for gathering information and field problems. To coordinate between researchers and the farmer who is actual user of technology it was felt that the facility of a 'single window' approach at the entrance of the ICAR Institute/State Agricultural Universities will enable the farmers to have the required information for the solution to their problems.

Thus the establishment of an Agricultural Technology Information Centre was made with view to provide such a mechanism beyond the individual unit of a research institution to contribute to the dissemination of the information. All the information on technologies, techniques, knowledge and materials, including planting materials, seeds, advisory services, diagnostic services, etc are supplied through a single window system.

The objectives of ATIC are:

- ✰ To provide and single window delivery system for product and species from an institution to farmers and other interested group as a process of innovativeness in technology dissemination at institute level.
- ✰ To facilitate farmer's access to the institutional resources available in terms of technology, advice, technology products *etc.* for reducing technology dissemination losses.
- ✰ To provide mechanism for feedback from the users to the institute.

Presently India has 44 ATIC. Such information will be useful for Farmers; Farmer-entrepreneurs; Extension workers and development agencies; NGOs; and Private sector organizations

7. National Agricultural Innovation Project (NAIP): 2000

The India National Agricultural Innovation Project works to facilitate an accelerated and sustainable transformation of the Indian Agriculture to relieve poverty and improve income through collaborative development and application of agricultural innovations by the public organizations in partnership with farmer's group.

The project has four components. Component 1 strengthens the Indian Council of Agricultural Research (ICAR) as the catalyzing agent for managing change in the Indian National Agricultural Research System (NARS) by focusing on: Information, communication and dissemination system; Business planning and development; Learning and capacity building; Policy and gender analysis and visioning; Remodeling financial management and procurement systems; and Project implementation.

Component 2 encourages research on Production to Consumption System (PCS) for raising productivity, profitability and competitiveness. Component 3 promotes research on Sustainable Rural Livelihood Security (SRLS). Component 4 supports Basic and Strategic Research to address challenge and problems of national and global importance. The idea is to enrich science to attain global leadership in science and technology development.

NAIP is likely to complete by December 2012.

Chapter 9

Role of ICAR, SAUs and NGOs in Rural Development

Role of Indian Council of Agricultural Research (ICAR) in Rural Development

ICAR formerly known as Imperial Council of Agricultural Research; it was established on 16 July 1929 as a registered society under the Societies Registration Act, 1860 in pursuance of the report of the Royal Commission on Agriculture. It is an autonomous organisation under the Department of Agricultural Research and Education, Ministry of Agriculture, Government of India. The ICAR has its headquarters at New Delhi.

India is a place where more than 70 per cent of people live in villages and their main source of livelihood is agriculture, animal husbandry etc. ICAR played pioneering role in initiation and spreading Green Revolution and subsequent developments in agriculture in India. ICAR is the apex body for coordinating, guiding and managing research and education in agriculture including horticulture, fisheries and animal sciences in the entire country. It is one of the largest national agricultural systems in the world with over 90 ICAR institutes and 45 agricultural universities spread across the country.

ICAR through its research and technology development has enabled the rural India to increase the production of food grains by 4 times, horticultural crops by 6 times, fish by 9 times (marine 5 times and inland 17 times), milk 6 times and eggs 27 times since 1950-51, thus making a visible impact on the national food and nutritional security.

It has played a major role in promoting excellence in higher education in agriculture and thereby helped in rural development.

Role of State Agricultural University (SAU) in Rural Development

The SAU is major partner in growth and development in agricultural research and education under National Agricultural Research System. There are 45 SAUs and majority of them are established in rural area so that they can evolve, promote and impart higher education to take up emerging challenges for the transformation of rural areas. It undertakes academic, research, extension and training programmes for the empowerment of the people with a view to ensure free flow of knowledge, techniques and technologies as a two–way process for better of rural life. It provides training facilities to personnel engaged in rural development in collaboration with the state government. It provides consultancy in all aspects of rural development. The State Universities undertake monitoring and evaluation of national and state-level development programmes, shares ideas and experiences in rural development work.

Role of Non-Government Organization (NGO) in Rural Development

The NGO's have been functioning in India for more than forty years. It can be seen everywhere *i.e.* rural and urban areas. Empowering communities became the key issue in their work. They are active in development of fields like agriculture, livestock, environment, water, human rights, women movements, sanitation, rural development and many more. For improving the quality of life rural communities they are playing vital role through sustainable development approach and giving real powers to the people for self governance.

NGO have been gaining visibility and recognition in development processes only over the past two decades. Over years, the roles of NGO have moved from being mere delivery agents of various services to undertake training, capacity building, entrepreneur development programmes. NGOs have also pumped in a lot of western money for development in our country. But still detailed study of their role and contribution in development are not available.

Instructional Technology

Chapter 10
Concept of Teaching and Learning Process

Teaching

Teaching is process of arranging situation in which the important topics that has to be learned are brought to the attention of learner, their interest are developed, desired are aroused and action are promoted. Teaching not merely informs the students but it transforms and stimulates them to make behavioural changes.

Learning

Learning can be defined as the relatively permanent change in individual's behaviour potentials as a result of experience or practice. It is a process of progressive behaviour.

Teaching and Learning Process in Home Science Extension

Teaching – learning process is heart of education. The fulfillment of aim and objectives of education depend upon the 'teaching – learning process'. It is planned interaction that promotes behavioural change that is not a result of maturation or coincidence. It is the most powerful instrument of education to bring about desired changes in the students. In teaching – learning process, the teacher, the learner, the subject matter, teaching aids etc are organized in a systematic way to attain some pre-determined goal.

Steps in Extension Teaching

There are six sequential steps of extension teaching: attention, interest, desire, conviction, action and satisfaction.

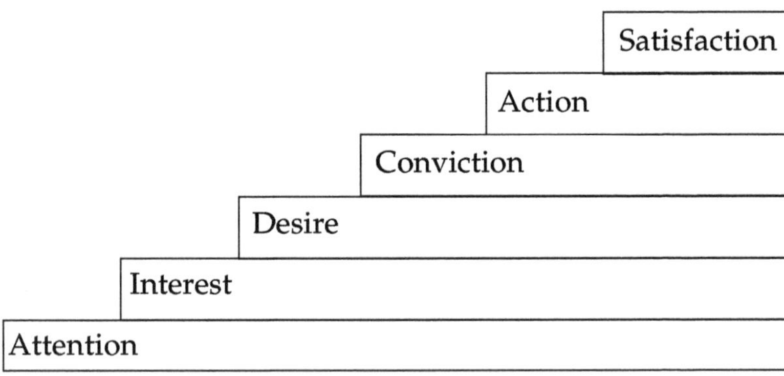

Figure 1: Steps in Extension Teaching

Attention

Attention is the starting point to arouse the interest. The first task of extension worker is to create awareness among the farmers/farm-women for innovations, latest scientific research and share the successful experiences of other farmers.

Interest

To arouse interest of the farmers or farm-women, the extension worker makes them realize the need or urge of innovation for further consideration. Presentation of one idea at a time in attractive manner is important factor in building interest.

Desire

Desire means motivating force sufficiently strong to built action. By supplying learner with all information on new idea or practice, the extension worker continuously stimulates learner's interest until the interest becomes a desire.

Conviction

It means convincing the learner for putting the new idea practice into action. This can be done by personal visit (home or farm) personal letter, telephonic conversation *etc.*

Action

After conviction, the extension worker should anticipate action. If action does not follow soon after desire and conviction then the new idea is lost.

Satisfaction

The end product of extension teaching is the satisfaction that comes as result of solving problem, meeting need, acquiring new skill or some other change in behaviour. Satisfaction can only be attained by learner, if they find progress in term of increased yield, income *etc.* At this stage extension worker should do follow-up and should also help learner to evaluate progress made.

Elements of Teaching – Learning Process/Learning Situation

Teacher/instructor, Learner, Subject Matter, Teaching aid and Physical facilities are the five elements of teaching learning process. Good teaching is well planned, sympathetic and co-operative activity.

To make teaching-learning effective, the physical facilities are well arranged in order to make noise free environment where the teacher teaches subject matter by using different teaching aid in an attempt to bring change in learner's knowledge, skill, attitude, experience and behaviour.

Elements of Teaching – Learning Process/Learning Situation

Sl.No.	*Elements of Teaching – Learning Process*	*Example*
1.	Teacher/Instructor	Extension worker or village level worker
2.	Learner	Villagers (farmer, farm women, rural youth, old age people etc.)
3.	Subject Matter	Any innovation (improved practice, new idea or new technology)
4.	Teaching aid	Chalk board, charts, posters, internet, radio, TV etc.
5.	Physical facilities	Sitting arrangement, ventilation, good visibility *etc.*

1. *Teacher/Instructor*: Teacher manages the teaching-learning process. He creates ideal situation so that proper learning can take place. Teacher can be extension worker or village level worker etc.
2. *Learner*: Learner is central element of teaching-learning process. In extension learner can be any villagers *i.e.* farmer, farm women, rural youth, old

age people etc. The learner are supposed to bring change in his behaviour through their own efforts like seeing keenly, listening carefully, practicing the learned things etc.

3. *Subject Matter*: It is any innovation such as improved practices, new idea, better technology etc. In extension the subject matter are selected based on need and interest of the villagers so as to make teaching-learning meaningful.
4. *Teaching aid*: It is means by which instructions are disseminated to the learner. It can be chalk board, charts, posters, internet, radio, TV etc. The use of teaching aids creates appropriate learning situation.
5. *Physical facility*: It includes sitting arrangement, ventilation, good visibility *etc* that makes the teaching-learning process effective and free from any noise. Noise any disturbance or distraction that hinders learning.

Effective Teaching – Learning Process

In extension the extension agent should proceed in following way so as to achieve effective teaching - learning process.

☆ Known to unknown	☆ Simple to complex	☆ Concrete to abstract
☆ Definite to indefinite	☆ Particular to general	☆ Induction to deduction
☆ Empirical to rational	☆ Psychological to logical	☆ Analysis to synthesis
☆ Whole to part	☆ Felt need to unfelt need	

Chapter 11

Classification of Extension Teaching Methods

The term 'instruction' means a 'deliberate arrangement of information' to help a learner achieve desirable change in performance and "technology" refers to systematic application of scientific or other organized knowledge to practical tasks.

So, Instructional Technology (Extension Teaching Methods) can be defined as systematic arrangement of information, so as to put the knowledge into practice in an effective manner to attain specific objectives.

Classification of Extension Teaching Methods

An extension teaching method is defined as sequence of progressive steps, undertaken to create situation that are conducive to effective learning. According to Legans (1961) extension teaching methods are the devices used to create situations in which communication can take place effectively between an instructor and the learner.

For an effective teaching – learning process instructor should be able to adopt each method or combination of methods effectively. They should be able to modify or improvise them according to the situation. To do this successfully an extension worker must know all methods available, when to use a given method and how to use it effectively.

The extension teaching methods can be classified on the basis of their use, form, learner type, objectives, stages of learning process, stages of adoption process, categories of adopters *etc.*

Classification of Communication Methods According to its Use

According to use of communication methods, it can be classified into: Individual, Group and Mass contact method.

1. Individual Contact Method

Here, the extension worker communicates with the people individually and maintains separate identity of each person *i.e.* it provides opportunities for face-to-face or person-to-person contact between the rural people and the extension workers. This method is followed when there are a few numbers of people to be contacted, the people are located close to the communicator and also sufficient time is available for communication. These methods are very effective in teaching new skills and creating goodwill between farmers and the extension workers. Example: Counseling, Farm and Home Visit, Telephone calls *etc.*

Advantages

- ☆ Helps to built rapport with the people since face to face interaction takes place
- ☆ Helps in gaining first-hand knowledge of farm and home
- ☆ Helps in selecting local leaders
- ☆ It is helpful in changing attitude of the people
- ☆ Helps in teaching complex practices
- ☆ It facilitates in getting feedback from the people
- ☆ It is flexible method

Limitations

- ☆ It is time consuming method
- ☆ It is relatively expensive method
- ☆ The coverage of people learners are low
- ☆ Biasness can occur towards some person

2. Group Contact Method

This method is adopted when the communication is to be taken place with number of people simultaneously, who have common interest and enough time available for communication. A group is defined as collection of small number of people within reciprocal communication. A group can be small, medium or large depending on number of people present. The size of small group may be from 15 to 25, medium group from 25 – 50 and large group from 50 – 100 persons. These methods also involve a face-to-face contact with the people and provide an opportunity for the exchange of ideas, for discussions on problems and technical recommendations and for deciding the future course of action. Example of group methods are: Lecture, Meeting, Result Demonstration, Method Demonstration *etc.*

Advantages

- ☆ It enables extension agent to have face-to-face contact with number of people at a time.
- ☆ It makes possible the selection of a part of target group.
- ☆ It strengthens the learning of the group since sharing of knowledge and experience can take place over here.
- ☆ It offers opportunities for interaction (between people and communicator) and feedback.
- ☆ It satisfies basic urge of people for social contacts.
- ☆ It is less expensive than individual method because of more coverage.

Limitations

- ☆ Difficulty can arouse in learning situation due to presence of wide diversity in interest of group members.
- ☆ Varied caste group and village fraction may hinder free interaction.

3. Mass Contact Method

It is characterized by communication with large heterogeneous audience. This method is used where the audience is large in number and is widely dispersed. These methods are more useful for making people aware of the new agricultural technology quickly.

Advantages

- ☆ It helps to create awareness amongst the people
- ☆ Here large number of people may communicate within short period of time

- It saves money and time
- It facilitates quick communication
- It also reinforces previous learning

Limitations

- It is less intensive method
- There is little scope for personal contact with audience and within the audience
- Generalized recommendation hinder application by individual
- There is difficulty in getting feedback

Classification of Communication Methods, Media and Aids According to its Use

Individual Contact	*Group Contact*	*Mass Contact*
1. Farm and Home visit	1. Lecture	1. Radio
2. Personal letter	2. Panel	2. Television and films
3. Telephone call	3. Symposium	3. Publications
4. Office visit	4. Discussion	4. Lecture recording
5. Office call	5. Tour (Field trip)	5. Posters
6. Counseling	6. Method of demonstration	6. Circular
7. Correspondence course	7. Slide show	7. Computer
	8. Training	8. Bulletin Board
	9. Flash card	9. Fair exhibition
	10. Drama	10. Campaign
	11. Songs	11. Mockups
	12. Speech	12. Dioramas
	13. Inquiry	13. Leaflet
	14. Forum	14. Folders

Contd...

Contd...

Individual Contact	*Group Contact*	*Mass Contact*
	15. Colloquy	15. Charts
	16. Clinic	16. News stories
	17. Institute	17. Pamphlets
	18. Workshop	18. Journals
	19. Seminar	
	20. Field day	
	21. Meeting	
	22. Contest	
	23. Interview	
	24. Clubs	
	25. Debate	
	26. Brain storming	
	27. Buzz session	
	28. Idea Inventory	
	29. Question period	

Classification of Communication Methods According to its Form

According to form of communication methods, it can be classified into: Written, Spoken, Visual and Audio-visual.

Classification of Communication Methods Media and Aids According to Form

Written	*Spoken/Audio*	*Visual*	*Audio-visual*
1. Bulletin	1. Meeting	1. Result demonstration	**Non-Projected**:
2. Leaflets	2. Farm and home visits	2. Exhibition	1. Method
3. Personal letter	3. Office call	3 Posters	2. Result Demonstration

Contd...

Contd...

Written	*Spoken/Audio*	*Visual*	*Audio-visual*
4. Journals	4. Telephone calls	4. Charts	3. Puppet show
5. News stories	5. Radio	5. Slides	4. Campaign
6. Circular letter	6. Recording	6. Film strips	5. Drama
	7. Lectures	7. Flash card	**Projected**:
	8. Panel	8. Flannel graph	Film
	9. Discussion	9. Bulletin boards	Video
			Television

Classification of Teaching Methods According to Functions

Telling	*Showing*	*Doing*
Farm and Home visit	Written words	Demonstration
Conference	Demonstration	Practical
Panel and forum	Tours	Trainings
Recordings	Pictures	Performance
	Photographs	
	Poster-charts	
	Exhibition	

Classification of Teaching Methods According to Stages of Learning

Interest	*Desire*	*Conviction*	*Action*	*Satisfaction*
Personal visit	Demon-stration	Result demonstration	New stories	Personal contact
Tours	Circular letter	Personal visits	Farm home visit	New stories
Demon-stration	Leaders	Friends and relatives	Local leaders	Field days
Video film	Meetings		Telephone call	
Meetings				

Classification of Teaching Methods According to Learning Objectives

Knowledge	*Skill*	*Attitude*
Printed matter	Demonstration	Training course
Lecture/Exhibitions	Case study	Demonstration
Discussion Planned	Supervised practice	Video films/T.V.
Tours	Workshops	Role playing
Case study		Puppet shows

Chapter 12

Extension Teaching Methods

Every extension teaching method cannot be used everywhere. Each method has its own use, importance and limitations. For an extension worker, it is important to know its concept, use, importance and limitations of each of the extension teaching methods so that they can use it efficiently.

Farm and Home Visit

It is a direct, face to face contact by the extension agent with the farmer, farm women at their farm or home itself. The aim of farm and home visit is to get confidence of the farmer and farm women so that they can gain first hand information on a topic. It is time consuming and costly method since only limited number of people can be contacted. This can also lead to biasness in extension work.

Discussions and Meetings

Discussion and Group meeting helps in democratically arriving at certain decision by a group of people, by taking into consideration the points of view of all the members'. It is a process that enhances people's participation in decision making and also develops capability of people to face challenges. Depending on the topic the number of members present in meeting can be called small (15-25) medium (25-50) or large (50-100) group meeting.

The aim of group meeting or discussion is to build a favourable climate for discussion for better understanding of the problem. It helps to pool new ideas generated within the group and it also helps information of opinion about a specific issue. It helps to develop favourable attitude and commitment for action.

It is slow process and thus may not be suitable in emergency when quick decisions are to be obtained in short period of time. Great skill is required to drive the discussion or group meeting in a desired direction.

Field Trip

Field trip provides unique opportunities for learning that are not available within the four walls of a classroom. A field trip is designed in a manner that participants can easily make connections between the objective of the field trip and the concepts they are learning in the rest of the educational program. Field trip for farmers are used to convince them and to provide them with an opportunity of seeing the results of new practices, demonstration skills, new implements etc. and to give them an idea regarding the suitability and application of these things in their own area. Good field trips provide participants with firsthand experience related to the topic or concept being discussed in the program. Field trip has following limitations *i.e.*

funding limitations, time constraints, and increased liability concerns.

Campaign

Campaign is an intense educational activity for attracting the attention of community towards solving a particular problem or to satisfy a need urgently felt by it. The duration of campaign varies from a day, few weeks or a few months; depending on the type of topic/problems to be solved.

The main aim of organizing a campaign is to create mass awareness about an important problem or felt need of community and also to encourage them to solve them.

Success of a campaign depends upon the co-operation of community and the leaders. For organizing a campaign requires adequate preparation, concentrated efforts and propaganda techniques and uninterrupted supply of critical inputs.

Exhibition

Exhibition is planned display of models, specimens, charts posters *etc.* in a sequence around a theme. The aim of exhibition is to create interest in wide range of people and promote visual literacy and motivate people to adopt better practices. Setting up exhibition, requires lots of fund and lots of preparation and planning.

Printed Materials

Printed material such as newspapers, journals, magazines, bulletins, letter, leaflets, folders, pamphlets, wall news-sheets and farm publications are used for communicating information to a large number of literate people.

1. *Journals/Magazines*: It is periodicals, containing information related to various topics of interest. It is not only for farmers but also for extension agents. It has fixed periodicity of publication.
2. *Newspaper*: Newspaper is bunch of loose printed papers properly folded, which contains news, views, advertisements *etc.* It is sold at regular intervals; mostly daily but few are sold weekly also.

   Extension agent cannot exercise only control over newspaper but, it can prove to be support fort extension work. It may support extension work by publishing news of extension activities, achievements, extension recommendation and package of practices, success stories, market news, focusing farmers' problems advertisements issued by extension organizations, inputs, dealers *etc.*

   The main limitation of newspaper is that it can target majorly literate population.
3. *Personal letter*: Personal letter is written by extension agent to particular farmer or farm women in connection with extension work. The aim of writing personal letter is to send information or to answer queries. It is useful in answering request for information as follow-up after visit, office call *etc.* with large farmers, e-mail is being used, instead of personal letter.
4. *Farm publication*: It is publication prepared by extension agency in printed form such as leaflet, folder, bulletin, newsletter, journals, magazines *etc.* It contains information relating to improvement of farm and home. The main aim of farm

publication is speedy mass contact at reasonably low cost. It provides support to other extension methods. It provides accurate information in written form that can be even used for future reference.

One of the limitation of farm publication is that only literate can take benefit from it. Its periodical revision is necessary to keep it up-to-date.

5. *Bulletin*: It is a bound booklet with number of pages containing comprehensive information printed on a topic.
6. *Newsletter*: It is mini-newspaper of good quality paper. It contains information of activities and achievements of an organization. It has fixed periodicity of publication.

Demonstrations

1. *Method demonstration*: It is relatively short time demonstration given before a group to show how to carry out an entirely new practice or an old practice in a better way. Here emphasis is given on effectiveness of carrying out a job. Example: Construction of improved chullah that will result in less of smoke production. The aim of method demonstration is to teach skills and stimulate people to action. It helps to improve upon the result by doing a job in a better way. One of limitation of method demonstration is that it needs good deal of preparation equipments and skill of the extension agent.
2. *Result Demonstration*: Result demonstration is method to show the value or superiority or worth

of a new practice by way of comparison with old one. The demonstrations are conducted in the farm or home of selected farmers. It is effective method for transfer of technology in community. Example: Worth of smokeless chullah compared to traditional chullahs. The aim of result demonstration is to show advantages and applicability of a new practice in farmer's own situation. It helps to built confidence of farmer so that they get motivated to adopt recommended practice. The limitation of result demonstration is that it needs more time, energy and fund for its execution.

3. *National Demonstration or composite demonstration*: It is method that shows use of many improved practices together. It is also called first line or front line demonstration. It is conducted by Researcher on farmer's field. National demonstration was first time conducted in India in 1965 by Ministry of Food and Agriculture.

Dramatization

Dramatization is an important technique through which student's knowledge on the character, situation or issue is increased in situation, issues and crisis. In this process verbal, written or visual language is enhanced. It can be performed as a classroom activity. Some of the examples of dramatization are: games, role-play, simulations and mime.

1. *Games*: Any game can be a means of learning. It has a set of clear rules. It includes an element of competition. Game tends to have winner or loser.
2. *Simulation*: The word simulation means imagination or replication. Here event or situation

made to resemble clinical practice. It involves application and integration of knowledge, skills and critical thinking. It is created to represent real environments. It deepens conceptual understanding of the real world. It brings change over time, reacting and adapting to your actions. It is free from the competitive element. It is excellent teaching strategy for many skills.

3. *Role-play*: It is most simplistic form. It includes individuals play themselves or another, placed in a given situation. Here person behaves as a result of what is going on around them. There is an element of game and simulation. It could have winners and losers. The environment clearly defined to simulate real world situation. It is an attempt by an individual to place themselves in the position of another or as themselves and deal with unfamiliar circumstances.

4. *Mime*: It is the art of teaching through nonverbal communication, through exaggerated expressions, actions and gestures. It is the art of portraying characters and acting out situations or a narrative by gestures and body movement. It helps in creating characterization *i.e.* students listen to instructions, watch arms, body movements. Here performers physically express their adopted characters by their facial expressions and in every body movement. Every movement has a motivation and purpose. It is a medium for performing roles using words and focusing on conveying richness of meaning through visual language alone.

Drawbacks of Dramatization

- ☆ The tutor gives up a high degree of control over the learning environment.
- ☆ Threat of simplification can result in failure to make optimum use of knowledge.
- ☆ Students can simply play role in a shallow and ineffective manner.
- ☆ A good and successful role-play depends upon the quality of the students involved and the seriousness with which they embrace the exercise.
- ☆ Students may see this as a frivolous and entertaining exercise.
- ☆ These can be time consuming both in preparation and playing.
- ☆ If included in formal assessment process this presents and additional problem of assessing student contribution and performance.

Factors Affecting the Selection of Extension Methods

Extension work requires numerous methods and teaching tools. The extension-teaching methods are the tools and techniques that help to create learning situations. Each has its place, and they supplement one another. Together they provide the stimulus for interest, desire, action, and achievement. It is necessary to properly select these methods for a particular situation. There are many factors affecting the selection of extension methods.

- ☆ The number of people extension agent wishes to contact *i.e.* individual, group or mass.
- ☆ Background of the people *i.e.* age, gender, educational level, economic condition, social believes etc.

- ☆ The skills of the extension workers in handling the teaching aid
- ☆ The content or subject matter to be disseminated *i.e.* knitting or some recipe can be taught through demonstration method and information about vaccination during pregnancy can be explained through lecture method.
- ☆ The combination of teaching aids to be used.
- ☆ Physical facilities available in the village such as electricity, meeting place etc
- ☆ Availability of finance for the programme
- ☆ Availability of time for the programme
- ☆ The length of time, the programme has been under way in a given area
- ☆ The stage of development of the programme in an area. For example: a result demonstration may no longer be needed to give proof of a recommended practice.

Chapter 13

Various Extension Teaching Aids

Audiovisual Aid

Visual aid can be seen but cannot be heard. Most people retain only 10-15 per cent of what they read. Audio aid is instructional device that can only be heard. It cannot be seen. About 20-25 per cent learning is retained if only audio aid is used for teaching. Audio-visual aid can be seen as well as heard. Up to 90 per cent of a lecture is retained if all senses are involved in learning.

Importance of Audiovisual Aids

- ☆ Makes understanding easy: because the learner can hear as well as can see
- ☆ Quick grasping: It helps to grasp complex method/ information fast

- Long lasting: Learning through audiovisual aid involves maximum senses therefore whatever is learned is remembered for long time.
- More effective: because the learner can also see the result/output of the improved practice
- Gives clear picture of the improved practice

Audio visual aid can be of two type *i.e.* Projected or Non-projected. Projected aid includes film, video and Television etc. while Non-projected aid includes demonstration, puppet show, campaign and drama etc.

Function of Audio-Visual Aid

- Helps to generate awareness
- Teaches complex method
- Saves time and energy
- Wide coverage can be obtained
- Repetition of programme for giving information about innovation can be easily done

Graphic Communication Media

Graphic are visual presentations on some surface, such as a wall, canvas, computer screen, paper or stone. Graphic aids are the tools of teaching through the sense of sight. It is used to inform, illustrate or entertain the people. Graphic communication is the process of creating, producing, and distributing material containing words and images to convey data and concept. Example: charts, poster, leaflet, folder, photographs, maps etc.

Charts

A chart is visual aid which explains the ideas, facts, and relationships on subject matter; with the help of graphics, pictures, numerical, comparisons, contrast, summaries and other helpful services in explaining the message. Charts are helpful in communicating difficult and dull subject matter in interesting and effective ways in extension.

Type of Charts

- ✰ *Simple chart*: In this type of chart information for only one topic is given in simple way. This type of chart is used by organization to give information about their scheme.
- ✰ *Bar chart*: Bar charts are made up of a series of labeled horizontal or vertical bars based on a measured scale. They are used to compare qualities at different time or under different circumstances. This is also called as a histogram.
- ✰ *Pie chart*: A pie chart is a type of a circle graph and used to show how several parts make up whole. The whole quantity depicts entire sample space and the pieces of pie in the circle graph are called sectors that depict part of the sample.
- ✰ *Tabular chart*: Tabular charts are used to bring together in compact form a mass of related data in table form.
- ✰ *Tree chart*: Tree charts are used to show development or growth of a thing. The origin is a single line or trunk and development is shown by branches of the tree.
- ✰ *Flow chart*: Flow charts are shown by lines and arrows. It helps to show organizational structure of any institution.

- *Pictorial chart*: Pictorial charts include pictures. Each visual symbol indicates quantity.
- *Overlay chart*: It contains a number of sheets that can be placed one over the other. A part of whole is drawn on each of individual sheet. This enables the viewer to see different parts at a time and also how they appear when one is placed above another. Ultimate product can be seen after the final overlay is placed. It provides dramatic view to the people.
- *Pull chart*: It consists of written matter on a large sheet of paper. The every single message is covered or hidden by piece of thick paper held in position. By pulling concealing strip one after another, message can be shown to the people. This provides suspense to the viewer.
- *Flip chart*: A number of charts are prepared on sub-topics of the main issue. Chart are arranged sequentially and bound together. For presentation the chart is arranged on stand and each chart is flipped one after another by the instructor. This helps is exposing a segment of the subject to the audience, thus is helpful in gaining attention.

Posters

A poster is a bold illustration with less of writing. It is boldly designed to attract attention towards an event, message, product or idea. It communicates only one idea at a time. Poster helps to generate awareness and interest among the people. It inspires and takes people towards action.

Flash Cards

Flash cards are a set of small compact cards with illustration on the front side and useful information on backside. The pictures on the theme are drawn on these cards in a logical sequence which is flashed before the audience. They are used to present step by step information. It helps to disseminate information on topics such as benefits of an innovation, social stigmas like dowry and violence etc. Flash cards enable the villagers to follow information more easily. The ideal size of flash card for a group of 30-50 people is considered to be 22 × 28 inch while for small group 11 × 14 inch is adequate.

Bulletin Board Display

A bulletin-board is board on which information is displayed. It is also known as tack board. It serves the purpose of making announcements in simple and local language, events and photographs of local activities are displayed.

Leaflet

It is single printed sheet of paper of small size, containing preliminary information relating to a topic. It has little illustration and can be printed on one side or two sides of a sheet. It is generally distributed free of cost.

Folder

It is single printed sheet of paper folded twice or four times. It consists of essential information relating to particular topic.

Pamphlet

An unbound printed work, usually with a paper cover. It may consist of a few pages that are folded in half and stapled at the crease to make a simple book.

Black-Boards

The chalkboard is most popular visual aid. Black-boards are most useful in group-teaching methods. They provide writing and drawing surface for chalk. Thus it is also known as chalkboard. They are frequently used in schools, colleges and meeting places. They make possible the use of sketches, drawings, words, symbols of a combination of them to emphasize a point. Unlimited flexibility and spontaneous modification is possible through blackboard teaching. One disadvantage is that it requires great preparation and a lot of creativity on the part of instructor. Olive coloured chalkboard with yellow chalk gives maximum visibility during evening hours.

Flannel-Graphs

Flannel graph is any kind of visual with piece of sandpaper or any adhesive material on the backside. They can be made to adhere easily to flannel board which is made by fixing a piece of thick flannel cloth on the board.

Projected Aids or Optically Projected Aids

Projected aids are those aids that require audio-visual equipment to be presented properly. Here a bright light is passed through a transparent picture, and by means of a lens, an enlarged picture is thrown or projected on the screen. Slides, filmstrips, transparencies and motion pictures are the examples of projected aid.

Transparencies

A transparency is a thin sheet of transparent flexible material, typically cellulose acetate, onto which figures can be drawn. These are then placed on an overhead projector or LCD projector for display to an audience. The standard size is (19 x 23 cm).

Slides and Slide Projector

Slides are one of the best known still projected aids. It is a transparent picture or photograph that is projected through a slide-projector which brings the enlarged image into focus on a screen. It is individually mounted.

Slide Projector or Diascope popularly known as Magic Lantern, is an optical aid to the process of teaching. It is used for projecting pictures from a transparent slide on a wall or screen. As it is used to project slides, therefore it is called a slide projector. It helps in showing the magnified image of the slide. The slide is placed inverted into the slide carrier part of the magic lantern (slide projector). The slide projector projects its erect image on the wall or screen by enlarging its dimension and making the vision more sharp and clear. The slide projector is useful for small as well as large groups.

Film Strips

It is an improvement upon slide projector (magic lantern). They are a series of pictures depicting a single idea, printed on a single length of strip of 35-mm film. Slides produced on films are called film strips. The device may be used as a slide projector or as a film strip projector. Instead of using different slide for different topics one strip or piece of still film is prepared. A film strip consists of a strip of cellulose acetate with 16mm or 35mm width and length 2 to 5 feet. It usually consists of 40 to 100 separate pictures related to a particular subject, topic or theme. Film strips can be shown to an large group of audience of about a 100 people. The additional advantage in using the film-strips is that the film can be stopped anytime during the show to explain or discuss a difficult or interesting point. A major disadvantage of filmstrips is that they cannot be repaired (spliced) easily.

Opaque Projector

The opaque projector can project opaque objects. The projector permits you to project an illustration or other material out of a book. It uses a 1,000-watt projection lamp and is equipped with a 4 1/2-inch diameter, 22-inch focal length lens. The projector comes with an opaque object-holding device and a conveyor belt loading mechanism that permits uninterrupted projection. It also has an optical pointer that permits details anywhere on the screen to be pointed out by a projected light spot.

The Overhead Projector

The name 'Overhead projector' comes from the fact that the projected image is behind and over the head of the speaker/teacher. In overhead projection, a transparent visual is placed on a horizontal stage on top of light source. The light passes through this transparency and then is reflected at 90° angle on the screen at the back of the speaker.

Electronic Media

Electronic media are media that use electronics for the audience to access the content *i.e.* it is mode of electronic transmission. It includes Radio, Television, e-mail and Internet (wide-open), Extranet (using Internet technology to link a business with information only accessible to collaborating parties)

Radio

Radio is an electronic audio-medium for broadcasting programmes to the audience. It is a mass medium of communication that can reach a large number of people at a given time involving the least expense. The major objective of radio for rural mass is to makes farmer aware of latest technologies and researches taking place in the world.

Television

Television is an electronic audio-visual medium which provide pictures with synchronized sound. The main aim of Television for rural mass is to create general awareness about agricultural and rural development programmes

Internet

The Internet is a global system of interconnected computer networks so as to serve billions of users worldwide. It is a network of networks that consists of millions of private, public, academic, business, and government networks, of local to global scope, that are linked by a broad array of electronic, wireless and optical networking technologies. The Internet is decentralized by design. Each Internet computer, called a host, is independent. Its operators can choose which Internet services to use and which local services to make available to the global Internet community.

E-mail

Electronic mail, commonly called email or e-mail, is a method of exchanging digital messages from one to one or more recipients. Modern email operates across the Internet or other computer network.. Today's email systems are based on a store-and-forward model. Email servers accept, forward, deliver and store messages. Neither the users nor their computers are required to be online simultaneously; they need connect only briefly, typically to an email server, for as long as it takes to send or receive messages.

Folk Media and Type

Folk media means traditional methods of information dissemination by rural people. It is method by which rural people communicate, inform or exchange their ideas, make social protest, educate society and entertain themselves. These include traditional theatre or drama, storytelling, song, dance, puppet show, proverbs and riddle/puzzles.

Importance of Folk Media

Folk media uses local language and dressing style. As folk promotes rural culture thus people find it more acceptable. Live performance makes folk more entertaining and informative.

Puppet Show

Here puppets (dolls) are dressed in traditional cloth. Synchronizing with the traditional song, music or dialogues in the background the puppets are moved on the stage with the help of string. Through this method message regarding health, agriculture, social issues etc. can be easily conveyed to large number of people in form of short story and brief dialogue

Three Dimensional Communication Aids

It includes all the audiovisual materials nearest to living experiences such as specimen, object and models.

Specimen

Objects which are incomplete or which are representative of a group or class of similar objects. For example- A laboratory specimen is an individual animal, part of an animal, plant, part of a plant, used as a representative to study the properties of the whole population of that species.

Objects

Real things which have been removed as units from their natural setting.

Models

Recognizable three-dimensional representation of real things. It creates a sense of realization in a person. For example, if a science teacher is presenting on the anatomy of human intestines, a model comprised of pieces replicating a human's inner organs can be held up and assembled to show students how they fit together and their relationship to each other.

Communication and Diffusion of Innovations

Chapter 14
Communication and its Barriers

Communication

The word 'communication' is derived from the Latin word 'communis' that means 'common'. Communication as defined by Legan (1961) 'process by which two or more people exchange ideas, facts, feelings or impression in a way to gain common understanding of the meaning, intended and use of message'. Communication is social, continuous and a two way process.

Importance of Communication

Communication is a central activity in most human and organizational activities. Without communication, there is no way to exchange thoughts, ideas and feelings. The communication becomes crucial when one need to fulfill a goal, network or spread some idea etc. Communication is said to be effective when it is simple and precise. The best way to start communication in extension is by identifying what need to be communicated. The importance of communication in extension can be summarized as follows:

1. Communication helps in socializing *i.e.* aid to share idea, opinions and recommendations.
2. It facilitates to convey directions and provide synchronization.
3. It helps to strengthen existing relationships between extension worker and the villagers and to form positive bonds for future interaction.
4. Communication helps to prevent mis-understandings
5. It assists in basic management functions such as planning and organization
6. It also assists in leading and controlling process.
7. Communication helps to motivate people for action by informing and clarifying them about the task to be performed and its outcome.
8. Communication is a source of information as it helps identifying and assessing alternative course of actions that helps in decision-making process.

Elements of Communication

There are six major elements of communication. These are:

1. *Communicator*: The communicator also known as source is the originator of the idea or message as it starts the process of communication. It is communicator who decides what messages are to

be sent, how to treat them, what channels to use and which receiver to reach. Communication is said to be successful if communicator skillfully communicates the message to the audience in order to get clear understanding and desirable action.

2. *Message*: Message is a content of information, instructions or orders passed across. It is information that a communicator wishes his audience to receive, understand, accept and act upon.
3. *Channel*: The channel is the medium through which the message is communicated. It is a physical bridge between the communicators and the receivers of the message. The channel must be able to transmit the message from one party to another without changing the content of the message. All extension teaching methods are the channels of communication. The channel can be a piece of paper or medium such as radio, email, television etc.
4. *Treatment of message*: Treatment of message means to encode the information/idea for transmitting the message to the receiver and later the receiver decodes the message for better understanding.

   Encoding means how the message is converted into a suitable form for transmission to another party. The medium of transmission determines the form of the communication such as the message takes different form if the communication is spoken or written.

   Decoding is the process where the message is interpreted for its content. It also means the receiver thinks about the message's content and internalizes the message. This step of the process is where the receiver compares the message to prior experiences or external stimuli.

5. *Audience*: They are the ultimate receiver of the message who decodes the message. For communication to be successful the receiver's need, knowledge level, attitude, available resources, customs and tradition need to be kept in mind.
6. *Audience response/feedback*: Feedback is the final step in the communications process. The feedback is the response of the receiver to the message which enables the communicator to carry out corrections or change message to be effective.

Barriers of Communication

Barrier in communication is hindrance in transmission of message. Obstruction or barrier in communication is also called Noise. Noise is any stimulus that interferes with a message being transmitted from a sender to receiver. Noise can be: External, Internal and Semantic.

- ✰ *External Noise*: Anything outside of a person that distracts the listener during communication such as sight, sound, smell, texture and environment. It includes things such as the sounds of road construction outside the window that make it difficult to hear what is being said.
- ✰ *Internal Noise*: It is mental interference *i.e.* thoughts or feelings that prevents you from listening. Such as day-dreaming, feeling tired etc.

- ☆ *Semantic Noise*: Semantic noise occurs when there is no shared meaning in a communication. It occurs when someone is dealing with medical professionals, lawyers, scientists or others who use terminology that lay man may not understand. It also includes reactions aroused unintentionally by symbols or words, causing the listener to abruptly focus on unrelated topics such as discriminatory remark, red flag words like taxes, cancer or controversial symbols.

An extension agent must discover barriers in communication so that they can take preventive steps to avoid them and make communication effective. Noise or barrier in communication can be on part of communicator, receiver, message or channel through which information flows.

- ☆ *Communicator as barrier*: If communicator fails to create favourable atmosphere for communication or fails to present the idea clearly to the audience; if he feels that the audience have always clearly understood; talks while others are not listening; fails to understand others point of view then the communicator is barrier in communication..
- ☆ *Message as barrier*: If message is not organized properly or it is not according to need or interest of the audience or inaccurate symbols, words, visuals or real objects is used to represents the message then the message becomes the barrier in communication.
- ☆ *Channel as barrier*: Channel becomes barrier in communication when it fails to reach the intended

audience due to inappropriate selection or improper handling of channel or failure in using enough channels in parallel or failure in avoiding physical distraction while message is conveyed.

- *Receiver as barrier*: Audience becomes barrier in communication when he lacks ability to think, listen, interpret and understand the message clearly. Heterogeneity of audience, lack of positive attitude towards the message conveyed or low knowledge level also befall the receiver as barrier in communication.

Ways to Overcome Communication Barriers

Barriers in communication can be overcome by following the general guidelines to facilitate communication.

1. Develop a positive attitude about communication in the audience.
2. Work at improving communication skills. Use simple words to convey the message for easy understanding
3. Make communication goal oriented. When the sender and receiver have a good relationship, they are much more likely to accomplish their communication goals. Therefore rapport building is essential for effective communication. Also showing empathy let the audience know that the communicator have sympathy and compassion towards them and this gives sign of genuine interest.
4. Approach communication as a creative process rather than simply part of the chore of working

with people. Learn the art of listening *i.e.* to hear and understand the questions and doubts of audience.

5. Vary channels to overcome channel barriers and feedback techniques.
6. Accept the reality of miscommunication. Therefore by randomly asking indirect questions, the communicator can get feedback.
7. Experiment with communication alternatives. What works with one person may not work well with another person.

Chapter 15
Diffusion of Innovation

Diffusion

Diffusion is process by which an innovation is communicated through certain channels overtime among members of social system. It is process of social change. Diffusion is special type of communication because it is concerned with new idea. The concept was first studied by the French sociologist Gabriel Tarde (1890).

Elements of Diffusion

In 1962 Everett Rogers, a professor of rural sociology published book 'Diffusion of Innovation' in which he proposed 4 main elements that influence the spread of a new idea: the innovation, communication channels, time, and a social system.

1. *Innovation*: Any idea, practice or object that is perceived as new by an individual, family, school,

company, office/organization is known as Innovation.

2. *Channel*: Channel is means by which message gets transferred from one individual to another. For example transfer of message can be through radio, television, newspaper, internet, phone, books, journals etc.
3. *Overtime*: Overtime refers to time taken by potential adopters in adopting an innovation.
4. *Social system*: Social system is defined as set of interrelated units that are engaged in joint problem solving to accomplish a common goal (Roger 1995). Social system constitutes a boundary within which innovation diffuses. Diffusion always occurs within social system. Norms are established behaviour patterns for the members of social system.

Characteristics of Innovation with Special Reference to Homestead Technologies

The term homestead means 'related to home'. 'Technology' means combination of knowledge, input and management practices which are used together with productive resources to gain a desired output. Technology represents mixture of hardware and software. Hardware component consists of tools, *i.e.*, part of machine like washing machine, flour mill, refrigerate, oven, cooker, vacuum cleaner. Software component consists of knowledge for tools or equipment *e.g.*, how to operate washing machine, how to preserve pickles, store woolen clothes *etc.*

Homestead technology means suitable technology to assist the home maker in performing jobs inside the house and managing the house in efficient and effective manner.

Characteristics or Attributes of Innovations

1. *Relative advantage*: It is degree to which an innovation is considered to be better than the idea it contained earlier. It can be economic advantage or technological advantage. *e.g.*, LPG Gas is technically advanced than Mud Chula so people are gradually adopting it.
2. *Compatibility*: It is degree to which an innovation is considered to be consistent with existing values, past experiences and needs of potential adopters. Compatibility has two dimensions: situational compatibility and cultural compatibility. Situations compatibility means if an innovation suits economic and climatic condition of people then adaptability is high such as washing machine will be adopted in an area only if sufficient water and continuous electricity is available. Cultural compatibility means if innovation is in agreement with their beliefs and values. For example toilet paper is not costly still it is not compatible to Indian believes and cultural values.
3. *Complexity*: It is degree to which an innovation is considered to be relatively difficult to understand and use. If an innovation is complex to use or understand its adoption is less. For example: Ovens or sewing machine are not very costly but due to its complicated procedures not all adopt it.

4. *Trialability*: It is degree to which an innovation may be experimented with, on limited basis *i.e.*, if innovation can be tried on small scale then its rate of adoption is higher. If small packets of a costly brand washing powder is available in market then women does try it and if convinced then they go for its adoption.
5. *Observability*: It is degree to which result of innovation are visible. It has two dimensions *i.e.* curative and preventive. As we all know that prevention is better than cure, still people hardly go for preventive measures. The reason being, the results of preventive measure are not visible but the results of curative measure are obvious and can be seen. For example medicine to cure swine-flu will be more adopted than the medicine that prevents from swine-flu.
6. *Predictability*: It is degree of uncertainty of receiving the benefit from adoption of innovation. If people can predict profitability by adoption of an innovation then the adoption rate is higher for it.

Stages of Adoption of Innovation

Adoption is a gradual process; it does not take at once. Adoption process starts from the very moment when an individual merely gets aware of a new idea. Then he undergoes a series of process that is called Innovation decision process or stages of adoption.

1. *Awareness*: Awareness occurs when individual is first exposed to innovation by chance or accidentally. Mass media plays great role during

awareness stage. At this stage where individual is aware of idea but lack detailed information about it. Thus individual wants to seek more information about it. It is also called knowledge stage. For example an individual comes across an innovation that there exists some type of *chullah* that produces less smoke.

2. *Interest*: At this stage, individual develops interest and seeks additional information about the innovation *i.e.* what it is, how it works, what is its importance. At interest stage, individual forms favourable or unfavourable attitude towards innovation. This stage is also called persuasion stage. Interpersonal communication play important role during interest stage to gather more information.
3. *Evaluation stage*: It is also called decision stage. Here individual makes mental application of new idea. Individual also decides whether to try or not to try the innovation. At evaluation stage individual judges worth of innovation.
4. *Trial implementation*: Trial stage occurs when individual puts an innovation into use *i.e.*, he actually applies or tests idea on small scale. It is also called implementation stage. Reinvention occurs at Trial Stage. Reinvention is degree to which an innovation is changed or modified by a user in the process of its adoption and implementation.
5. *Adoption*: When individual uses the new idea continuously on full scale then it is called adoption.

At adoption stage individual takes final decision to apply an innovation on continuous basis. It is also called confirmation stage.

Adopter Categories

Every individual does not adopt an innovation at the same time. Few adopts early and some in later span of time. Rogers defines an adopter category as a classification of individuals within a social system on the basis of innovativeness. In the book *Diffusion of Innovations,* Rogers suggests a total of five categories of adopters. The adoption of an innovation follows an S curve when plotted over a length of time.

Categories of Adopter

The categories of adopters are: innovators, early adopters, early majority, late majority, and laggards.

1. Innovator

- ☆ They are first to adopt new idea
- ☆ They are Venturesome and are willing to take risk
- ☆ They are young, active and very eager to try new idea
- ☆ They are merely 2.5 per cent in number
- ☆ They are cosmopolite, have good contact to scientific sources and wide social relationship
- ☆ They are financially sound to cope with uncertainty about success of innovation

2. Early Adopter

- ☆ They do not test untried idea but are quick to use tried idea

- ☆ They are Respectable and have the highest degree of opinion leadership
- ☆ They are generally 13.5 per cent in social system
- ☆ They serve as role model
- ☆ They are younger in age, socially active, have more financial lucidity, advanced education, and are more socially forward in comparison to late adopters.
- ☆ They are judicious in choosing an innovation for adoption
- ☆ They are individual 'to check with' before using new idea
- ☆ Extension workers contact them for spread of innovation

3. Early Majority

- ☆ They adopt an innovation before average members of social system but their time of adoption is significantly longer than the innovators and early adopters
- ☆ They are deliberate *i.e.* think carefully before adoption of innovation
- ☆ They are generally 34 per cent in a social system
- ☆ They are not very early nor too late to adopt an innovation
- ☆ They have average education, economic and social status
- ☆ They hardly hold leadership position

4. Late Majority

- ☆ Individuals in this category will adopt an innovation after the average member of the society.

Chapter 17
Factors Affecting Adoption

There are many factors that affect adoption of innovation; these can be personal, communicational, cultural or sometimes situational factors.

Personal factors: such as age, education, knowledge level etc. plays important role in adoption of an innovation. Older farmers have lower rate of adoption, because of their traditional nature, in comparison to younger farmer. Educated farmer have higher rate of adoption in comparison to illiterate farmer; the reason is that educated farmer have positive attitude towards new technology.

Communicational factors: such as social interaction, mass media exposure affect adoption of innovation. The higher the farmer is cosmopolite, the more open he will be new ideas and new technologies thus higher will be rate of adoption. Similarly cultural factors such as openness to new idea and overcoming the traditional barriers also affect adoption.

Situational factors: such as income of individual, supply of goods and services, government policies etc. affect adoption of new idea. If all the above stated factors are overcome but the farmer does not have sound economic status then adoption of innovation cannot take place.

The rate of adoption can be easily understood from table given below.

Factors Affecting Adoption of Innovation	*Rate of Adoption*	*Reason*
Age: Older farmers	Lower is the rate of adoption	Their traditional nature stop them from adopting innovation
Education: Educated farmer	Higher is the rate of adoption	They have positive attitude towards new technology
Communication: Cosmopolite Farmer	Higher is the rate of adoption	They are more open towards new ideas and new technologies
Income: Rich Farmer	Higher is the rate of adoption	They have high risk bearing capacity

Type of innovation-decision such as Optional, Collective, Authority and Contingent decision also affects the rate of adoption of an innovation.

Type of innovation-decision	*Decision Power*	*Rate of Adoption*	*Example*
Optional	Choice to adopt/ reject an innovation by an individual is independent of decisions of other members of social system	Faster	Decision to buy bed-sheet, toothpaste or towel for own home

Contd...

Contd...

Type of innovation-decision	*Decision Power*	*Rate of Adoption*	*Example*
Collective	Choice to adopt/ reject an innovation that are made by consensus among the members of a system	Slower	Decision for buying car, computer or washing machine at home
Authority	Choice to adopt or reject an innovation that are made by a relatively few individuals in a system who possess power, status or technical expertise.	Faster	Decision for type of dress in school or college as stated by Principal or Dean of the University
Contingent	Choice to adopt or reject that can be made only after a prior innovation decision. Such as, an individual member of a social system may be free to adopt or not to adopt a new idea only after this system's innovation decision.	Slower	A Professor may need multimedia for teaching but, it depends on decision of authority who will decide to purchase multimedia.

Chapter 18

Consequences of Adoption of Innovation

Consequences mean change that occurs within social system as result of adoption of innovation. No matter the decision is Optional, Collective, Authority or Contingent; every decision brings change in social system and these changes can bring positive or negative consequences or results.

Positive Consequences	*Negative Consequences*
Desirable: It means consequences that are desired by individual. These consequences are functional that means the change are useful for individual For example: Tractor helps to increase yield	**Undesirable:** It means consequences that are not desirable. These consequences are dysfunctional that means the change are not needed by an individual For example: Tractor creates Pollution

Contd...

Contd...

Positive Consequences	*Negative Consequences*
Direct: Change that occur in social system due to immediate response of innovation (Positive change) For example: Smokeless chullah gives less smoke	**Indirect:** Change that occur due to direct consequences. For example: Due to hurry improper smokeless chullahs were constructed that led to fire in the cottage
Anticipated or Manifest consequences: Change that are recognized and intended by members of social system For example: Electricity in village	**Unanticipated or Latent consequences:** Change that are neither intended nor recognized by member of social system. For example: Need of cyber in village.

Rejection

Rejection is decision not to adopt an innovation. It is opposite of adoption.

- ☆ *Active rejection*: Person first implements an innovation but later decides not to adopt it.
- ☆ *Passive rejection*: When individual never considers an innovation to implement then it is called Passive rejection. It is also called non-adoption

Discontinuation

Discontinuation is decision to reject an innovation after having previously adopted it. It can be Replacement discontinuation or Disenchantment discontinuation.

- ☆ Replacement discontinuation: It is decision to reject an idea in order to adopt better innovation.
- ☆ Disenchantment discontinuation: Disenchantment discontinuation is decision to reject an idea due to dissatisfaction.

Innovativeness

It is degree to which an individual or adopter is relatively early in adopting new idea, in comparison to other members of social system.

Innovation-Decision

Innovation-decision is process that consists of series of action or choice over time through which individual or organization undergoes for adoption of an innovation. It is the length of time required to pass through the innovation-decision process

Rate of Adoption

Rate of Adoption is speed with which an innovation is adopted by members of social system. It is measured by length of time required for certain percentage of members of a system to adopt an innovation. Rate of adoption is measured by using an innovation. The rates of adoption for innovations are determined by an individual's adopter category. In general individuals who first adopt an innovation require a shorter adoption period (adoption process) than late adopters. Most innovations have "S" shaped rate of adoption.

Over Adoption

Continuous/vigorous adoption of an innovation when expert thinks that it should not be done. *e.g.,* Over use of fertilizer/pesticide by farmers to increase yield of crop, though the experts thinks that it should not be used because it is harmful for health of the people.

Diffusion Effect

Diffusion effect is increasing pressure of interpersonal network for adoption of an innovation. These can be pressure by opinion leaders, change agent or Aide.

Opinion Leaders

Those who are able to influence other individual's attitudes informally in a desired way with relative frequency are called opinion leaders. It is type of informal leadership. Opinion leaders are centre of interpersonal communication network. They don't have any degree from the University. They are individuals within social system *e.g.,* Pearson next door to whom majority of people go for asking his opinion on any of their problem.

Change Agent

They are individual who influence people's innovation decision in a direction as stated by change agency. They are professionals with the University degree in technical field. *e.g.,* Extension Officer; Agriculture Officer.

Aide

They are less than fully professional change agent. They are more homophilous than change agent. *e.g.,* field worker (from village itself who has been given some training).

Programme Planning, Implementation and Evaluation

Chapter 19

Basics of Programme Planning

The Program planning is essential for extension professionals to build up programmes for rural development. It is flexible, sequential process to make decisions for proceedings of a programme. It helps to develop plan of action for what to do and how to set priority in a programme to achieve goal.

Programme

The dictionary meaning of the word "Programme" is a proclamation, a prospectus, a list of events, a plan of procedures, a course of action prepared or announced before hand, a logical sequence of operations to be performed in solving a problem.

Extension Programme

According to Kelsey and Hearne (1949) Extension

Programme is a statement of situation, objectives, problems and solutions.

Legans (1961) explicated that an extension programme is a set of clearly defined, consciously conceived objectives or ends derived from an adequate analysis of the situation, which are to be achieved through extension teaching activity.

In short an extension programme can be defined as: A written statement of situation, objectives, problems and solutions which has been prepared on the basis of an adequate and systematic planning effort and which form the basis of extension teaching activities in a specific area, for a given period.

Planning

Planning is a process which involves studying the past and present in order to forecast the future and in the light of that forecast determining the goals to be achieved and what must be done to achieve them.

Myerson and Banifield (1935) defined Planning as; designing a course of action to achieve ends.

Programme Planning

Programme planning is a decision making process. For facilitating development of community and individual, programme planning involves critical analysis of the existing situation and the problems, evaluation of the various alternative solutions to the problems and selection of the relevant ones, giving necessary priorities based upon local needs and resources available through cooperative efforts of the people.

Extension Programme Planning

Musgrow (1962) defined Extension programme planning as a process of determining, developing and executing programmes. It is continuous process, whereby farm people, with the guidance and leadership of extension personnel, attempt to determine, analyze and solve local problems.

The definition comprises of three questions:

- What needs to be done?
- When it should be done?
- How it should be done?

In other words extension programme planning can be defined as: A decision-making process in which extension educationists involve people's representatives, to determine their needs, problems, resources and priorities in order to decide on an extension programme consisting of situation, analysis, problems, objectives and solutions, which will form the basis of extension teaching plans for a given period.

In short it implies that extension programme planning is:

- A process of decision making,
- Involves advance thinking,
- A progressive step by step process,
- Involves people in defining the goals to be achieved,
- Uses educational means in defining the goals and situations,
- Built around improved technology, people, their needs, interests, resources, values, attitudes and skills,

- A written statement of situation, problems, objectives and solutions.

Few Concepts in Programme Planning

Process

It is the act of preceding a series of actions or operations definitely conducting to a desired end.

Situation

It is a statement of affairs that includes the cultural, social, economical and physical conditions in which a particular group of people find themselves at a given period of time.

Analysis

It is the process of breaking down or separation of whole into its components for studying the relationship of each component with each other. In other words, it is a process of looking anything closely, critically for having its complete understanding.

Need

Need is the difference between what is and what ought to be. In other words, need is lack of something that present, may further lead to welfare of an individual.

Interest

Interest means any aim or object which stimulates activity towards its attainment. It is activities for which we have liking or disliking and one which we wish to continue or discontinue.

Aim

Aims are generalized and broad statement of directions, with respect to given activities. It directs the intension or purpose for *e.g.*, to aim at distinction, to aim to do well in exam.

Objectives

Objectives are expression of the ends towards which our efforts are directed. It means a "direction of movement" or 'goal intended to be attained by an individual'. Objectives are always expressed in specific term. It is always accompanied by a time frame.

Goal

A goal is the outcome of a series of successfully completed objectives Goal may be defined as any given direction one expects to go during a given period of time. It is expression of intended outcome in general term.

Difference between Aim and Objective

Point of Difference	*Objective*	*Aim*
Exactness	Always as specific	Slightly vague or considered as a general statement
Time frame	Always accompanied by a time frame	Has no time frame
Duration	An objective is a (relatively) shorter term goal	An aim is a (relatively) long term goal

Problem

It is a condition or state of dissatisfaction of people's wants/desire.

Solution

It is a course of proposed action to change an unsatisfactory condition to one that is more satisfying.

Project

A project is an outline of procedure pertaining to some phase of extension work.

Plan

A plan is predetermined course of action.

Plan of Work

Plan of work is an outline of activities so arranged as to enable efficient execution of the programme. It is a written statement of activities to be undertaken by an individual, a group of people or an organization, within a definitely stated time, to carry out the recommendations in the programme. The plan of work indicates what is to be done, who is to do it how it is to be done, when it is to be done, who are to be served or reached and how the results will be measured. The plan of work includes "Calendar of activities."

A Calendar of Work

It is a plan of activities to be undertaken in a particular time sequence.

Decision Making

It is the process of consciously choosing courses of action from available alternatives and integrating them for the purpose of achieving a desired objective.

Principle

A principle is a statement of policy to guide decision and actions in a consistent manner.

Evaluation

Evaluation is process of determining extend to which objectives have been attained. It is about using information derived from monitoring to make judgments or conclusion that will aid/help in decision making.

Evaluation as applied to the field of extension education may be defined as "a process of systematic appraisal by which we determine the value, worth or meaning of an activity of an enterprise." It is a method for determining how far an activity has progressed and how much further, it should be carried out to accomplish objectives. Thus to an extension worker evaluation means determining the results of this extension programmes in order to know the extent to which objectives have been achieved and why and what changes would be needed in case the programme is planned again or in its implementation.

Monitoring

It is a management process for a continuous review of the flow of inputs, outcomes of outputs and for getting feedback for effective management decision. Monitoring keeps track of the project and progress.

Leader

A leader is a person who influences people through his behaviour and initiates action in the group.

Change Agent

A change agent is an individual who influences clients innovation decisions in a direction deemed desirable by a change agency.

Opinion Leader

Opinion leader is the person whom people contact for his opinion regarding any problem or an issue.

Opinion Leadership

It is the degree to which an individual is able to

influence other individual's attitudes or overt behaviour informally in a desired way with relative frequency.

Organization

An Organization is human grouping or social unit, purposely constructed to seek specific goals.

Voluntary Organization

According to Prabhakaran (1992) "A voluntary organization is an agency which works for the welfare of the community in any given area of its own volition. It may be just an individual or it may have a more formal structure. It is actually a group of well trained, committed persons having an area of activity dealing and interacting with people."

Objectives of having a Programme

The general objective of having an extension programme is to influence people to make desirable change in their way of life and in their style of living.

Kelsey and Hearne (1967) have given the following specific reasons for having a programme.

1. To ensure careful consideration of what is to be done and why? Means it gives direction.
2. To provide/furnish a guide against which to judge all new proposals. Because progress requires a design. Success results from design not from drift or trial and error.
3. To establish objectives towards which progress can be measured and evaluated.
4. To have a means of choosing the important (deep rooted) from incidental (minor, less important)

problems and the permanent from the temporary changes.

5. To develop a common understanding about the means and ends between various functionaries and organizations.
6. To ensure continuity during changes in personnel.
7. To help in development of leadership qualities.
8. To avoid waste of time and money and promote general efficiency.
9. To justify expenditure and to ensure flow of funds.
10. To have available in written form a statement for general public use.
11. To ensure better and efficient utilization of resources, accountability and human development.
12. It stimulates action keeping in view, needs, interests, priorities, resources or people in mind.

Characteristics of Programme Planning

1. *Planning process can be either simple or complex*: Some planning is simple and few others are complex. For example, planning for a lecture is simple while planning for socio-economic development is complex.
2. *Planning is a process of making decision*: Means, in planning process we decide objectives of programme, its significance and rationality. In other words, we develop a blueprint for action.
3. *Planning has a purpose*: Without some expressed purpose, there can be no planning. It aims at human development through the optimum use of

resources and the rational integration of community.

4. *Planning continuous process*: Planning is a continuous process because, it involves series of actions. It is never complete, because after completion of one step, situation changes which require further planning. On the same line, when people's simple problems are solved, they tend to identify more critical problems to solve.
5. *Planning is a progressive step-by-step process*: Planning has definite sequence. End of one step leads to initiation of next step and so on thereby the work progresses.
6. *Defining of goals is a basic part of planning*: Planning is always done to achieve some goals. In this process of goal setting, people are also involved. Planners help people in defining their goals.
7. *Planning is a social process*: Planning is a social process because it involves human interactions. Interaction assumes some type of communication between people thus planning can be said as social process.
8. *Programme planning is a social planning*: Social planning has been defined as a conscious, collaborative, interactional process combining investigation, discussion, agreement and action in order to achieve the conditions, relationships and desirable values. Individual and societal values are involved in social planning.
9. *Programme planning is an educational process*: Programme planning is an educational process as

it changes attitude of the people. It involves both teaching and learning process in which it teaches skills and gives knowledge to the people.

10. *Programme planning is a coordinating, unifying and integrating process*: Planning requires integrated and coordinated efforts of officials and non-officials, institution and organization, men and materials for success of the programme.
11. *Programme planning is an evaluating process*: Evaluation of the results after accomplishments of the programme is not the only criterion of programme planning but evaluation of planning steps are also done.
12. *Programme planning requires skills and abilities on the part of the planners*: A number of high level professional skills and abilities are required for planning of a programme, such as:
    - ☆ Understanding of the nature and roles of extension education organization.
    - ☆ Knowledge and understanding of the technology related to the subject with which the programme is concerned.
    - ☆ Ability to clarify the objectives of a programme and to state them that they are useful in guiding its execution.
    - ☆ Skill at seeing the relationship between principles and practice.
    - ☆ Skill at inquiry and human relationships.
13. *Programme planning requires advance thinking*: Planning requires knowledge, imagination and

reasoning ability. One needs to know "where we are" and "where we ought to go," and one has to judge "what to do" and "how to do." It is a complex exercise as it involves people, their needs, their interests, analysis of situation and making decisions about what should be done, determining useful actions, projecting the desired shape of things in future and several other components, which are rarely simple and thus require intense advanced thinking.

14. *Programme planning is built around content*: Without some expressed purpose, there can be no planning. Extension programme planning is built around available improved technology, the people, their resources, problems, needs and interests.
15. *Programme planning is a system*: It is a system as its procedures and processes are interrelated, ordered and linked progressively to form a collective whole. It includes several sub-processes such as planning designing, implementing, evaluation, *etc.* Its end products are extension programme.

Chapter 20
Principles of Programme Planning

Principles of Programme Planning

Several principles on programme planning have been given by various scientists but Sandhu (1965) identified a set of principles that may be more applicable in developing countries like India. The following principles should be followed while planning an extension programme.

1. *Extension programme planning should be based on analysis of facts in the situation*: Culture difference means there is variation in customs, values, attitude and way of living of the people. Any programme or technique cannot achieve the desired results, if it is not in harmony with the culture of the people because rural people have high faith in their traditional customs and values. Thus before

programme development one need to collect adequate information about the people, land type, crops grown, social structure, economic status of people, their communication pattern, habits, traditions, indigenous knowledge and the culture.

2. *Extension programme planning should select problems based on people's interests and needs*: The ultimate objective of programme planning is to satisfy the needs of the people. So those problem should be selected which are most urgent and of widest concern. Priority should be given first to the felt needs of the people. For effective extension, work must begin with the interest of the people to solve their felt needs in accordance to the resources available to them.
3. *Extension programme planning should determine definite objectives and solutions which offer satisfaction*: In order to hold interest of rural people in extension programme, the planner should set working objectives and offer solutions which are within their reach and gives satisfaction on achievement of selected objectives. Objectives should motivate people for action. The objectives must be periodically reviewed and revised to move towards progress.
4. *Extension programme planning should have permanence with flexibility*: An extension programme must be kept flexible to meet the changing needs and interests of the people and to adjust the changes aroused due to unforeseen circumstances. Rigidity in planning must be avoided because it is simply impossible to eliminate

the uncertainty of the future. So an extension programme should be prepared well in advance of its execution but not too far ahead of time. Permanence in programme means that it should be able to meet long term and short term goals as well as it should be able to overcome unforeseen circumstances.

5. *Extension programme planning should have balance with emphasis*: For programme to be successful it must be comprehensive enough to maintain balance among all age groups, creeds and races at all levels and community, block, state, national and international problems. Along with maintaining balance, the programme should be able to emphasize on solutions to recent problems and urgent needs.
6. *Extension programme planning has a definite plan of work*: No matter how well a programme is thought through, it is of no use unless carried out well. Therefore programme planning needs good organization and careful planning for action. A good plan of work is an outline of procedure, so arranged as to enable efficient execution of the entire programme thus it should answer to questions such as: what, where, when and how the job will be done.
7. *Extension programme planning is an educational process*: Extension programme is an educational process as it develops skill in the rural people to identify and solving their problems. People are involved in the programme, right from planning

process so that they may consider it as their programme. So, efforts should be made to involve a large number of people in identifying their needs and significant interests. Effective programme planning is a scientific, problem solving process, in which skilled thinking is necessary to help people and to overcome the complex problems of today's society. People should be involved in local surveys and neighborhood observations. It provides the opportunity to learn more about their own community, area and increases their interest. The extension worker should provide local leaders with the knowledge, skills and attitudes which are educationally helpful for people.

8. *Extension programme planning is a continuous process*: Education is a continuous process and extension programme is educational in nature. Hence, it can be said extension programme planning is a continuous educational process. Moreover new knowledge never exhaust in the field in which extension people work, due to constant flux of homestead technologies and teaching technologies.

   Extension people must make periodic adjustments in their plans to meet the changing problems. Extension people must also be alert to advancements in science and technology. With emergence of new technology, solutions to problems change. Therefore, it is necessary to view extension programme planning as a continuous process as its recurrence is cyclic.

9. *Extension programme planning is a coordinating process*: Planning must be coordinated within a program unit and collaboration must be evident between program units. Such coordination helps direct all of the needed and available resources toward achieving objectives. Those persons who are accountable for implementing plans should be involved in the planning. Planners must also coordinate the efforts of all well-informed and eminent leaders and groups, related agencies, government and private organizations etc. Above all they should obtain the interest and cooperation of the local people so as to get combined effort of all for development of the community.
10. *Extension programme planning involves local people and their situations*: Extension programme planning should involve local people at every step and the extension work should start from local situation and available research. People become interested and give better support to the programme if they are involved right from the planning process. So extension programmes should be planned with the people and not for the people.
11. *Extension programme planning provides for evaluation of results*: Extension programme planning involves decision making procedures, so evaluation is important in order to make intelligent decisions. Mathews (1962) pointed out that extension programme planning and evaluation go together. Kelsey and Hearne (1949) said that all principles of programme building are related to evaluation.

Effective evaluation depends on clear objectives. Starting a programme with the intention of having a careful evaluation at the end of a programme has a salutary effect upon the intermediate processes. So provision for both types of evaluation *i.e.,* concurrent and Ex-post facto types evaluation should be made in extension programme planning.

Some other principles that should be considered while planning for the programme are as follows:

12. *Extension programme planning should consist of clear achievable objective*: The goals should be effectively converted into specific objectives. The clear presentation of specific objectives that are measurable contribute toward the achievement of program goals is basic to all planning.
13. *Extension programme planning should have well defined time frame*: The extension programme planning must have a time frame in excess of one year if major programs are to be attempted.
14. *Extension programme planning should be revised from time to time*: Long-range plans should be reviewed and changed as necessary, at least once on an annual basis. Constant changes in society are such that planning over an extended time is periodically affected by unforeseen events. Consequently, a comprehensive annual review of goals, objectives and strategies, in the light of existing conditions and changes, is extremely important.
15. *Extension programme planning should be simple*: The planning process and the plans made should be kept simple and so designed, that the average staff

person in the programme can understand the programme, its objectives and implementation procedures.

16. *Extension programme planning should itself be well planned*: The process of program planning itself should be planned. It should fully involve the local people, extension workers, representatives from various coordinating institutions and organizations in every decision making process.
17. *Extension programme planning should conduct SWOT analysis*: Strength, weaknesses, opportunity and threat should be conducted well in advance. Before planning relevant facts should be found and keen analysis should be done planning involves decision-making. Experience has indicated that the quality of decisions cannot consistently rise above the quality of information on which such decisions are made. Consequently, the information for planning must include an analysis of the internal (extension organization) and external (rural) problems, needs, opportunities, strengths, and weaknesses.

Chapter 21
Steps in Programme Planning

Steps in Programme Planning

Programme planning is a step-by-step procedure. It involves decision making at every stage so as to attain the desired goal. The sequential steps for programme planning are given below.

1. Information Collection
2. Situation Analysis
3. Need Assessment
4. Setting up priority
5. Determining on objectives
6. Development of plan of work
7. Execution of plan
8. Evaluation of progress
9. Reconsideration

1. Information Collection

Success of any programme lies in establishment of strong foundation for it. This can be achieved by pre data collection of detailed information about local situation, resources available, local problems, culture and social status of local people etc. The facts secured from sources outside the community such as; from revenue records, Panchayat Samiti records, local newspaper or magazines, by personal visits, records of demonstration surveys, reports from key leaders, conferences and meetings and other sources are also valuable for success of the programme.

2. Situation Analysis

After collection of all local information, these facts should be transacted into familiar language so that it can be used for individuals or committees development. Extension Personnel should analyze these factors on the basis of "What they are" and "What they ought to be". Only after analyzing all these problems the solution should be considered.

3. Need Assessment

Extension programme is to be guided by societal needs. Analysis of the information collected help in assessing the need of the local people. A change agent by analysis shows the situation as it is and identifies the need with the help of extension worker. Need assessment helps to find local goal and priorities.

4. Setting Up Priority

After identification of the problem, the priority should be set to first solve, the most urgent and foremost problem, followed by next important problem and so on. The priority setting should be done by keeping in mind the felt need of the people; for which local leaders should be consulted.

5. Determining Objectives

After deciding upon the problem to be solved, its objectives should be developed. Objective describes the intended result on the target area and it must be measurable. The success and failure of a particular programme is judged in terms of whether or not objective is reached. Therefore while drafting the objectives the Extension Personnel should keep in mind the following points:

- ☆ The objective should be clearly stated
- ☆ Objectives should be limited and extremely important.
- ☆ Participation of rural people should be compulsorily included
- ☆ Objectives should be able to bring about desired change in human behaviour.
- ☆ The objective should be achievable considering the people concerned and available resources
- ☆ The objective should be developmental and lead to an expansion of interest and satisfaction.
- ☆ The objective should be such that they can be evaluated and evidences of accomplishments can be identified.

6. Development of Plan of Work

Plan of work is definite outline of procedure for solving the different problems of the extension programme. The plan of work should answer the following five questions:

(*a*) What different things or jobs need to be done to accomplish objectives (*b*) How best to do each of specific jobs

(*c*) Who will be responsible for planning, preparing and execution

(*d*) When each specific part of plan including meetings to be held

(*e*) Why things need to be done.

7. Execute Plan

Executing the plan means carrying through the previously planned work. The programme execution should be coordinated within the extension service and other agencies and organizations. The calendar of activities and events should be followed as planned. A variety of appropriate techniques, methods and materials should be used, as planned. Adequate and appropriate training and other assistance as planned should be given to lay leaders. The execution of plan should be carried out as decided and unnecessary changes should not be made.

8. Evaluation of Progress

Evaluation is process of determining the extent to which objectives have been attained. It is possible to evaluate an educational programme after it has been completed. But it is a much more difficult task. Therefore evaluation should be built into the programme from the start. It can be used at any step in the programme planning and implementation process. Impact of evaluation is used to determine if a program met the goals and objectives, identify program impact on individuals and communities, provide documentation for stakeholders, and meet accountability requirements. From evaluation it can be figured out whether programme is going in right direction or not.

9. Reconsideration

The findings of evaluation need to be used by applying it to future work or if evaluation results indicate that aim and objectives are not fulfilled and that work has to be done again than the whole programme is repeated from first step to the last step.

Evaluation process has not served its purpose unless conclusion is drawn from it has entered the continuing plans of workers.

Involvement of Local People in Planning Programme

Participatory planning means involvement of people in problem identification, prioritization and plan preparation. Participation promotes a sense of transparency and accountability among the people. People's participation is of crucial importance for the success of an extension programme. Involvement of people right from planning stage increases their participation also during implementation of the programme as they start feeling it as their programme. Participation has power to convert centralized planning system to decentralized planning system.

Factors influencing participation of rural people:

- ☆ The extent to which a programme is based on the felt needs of the people.
- ☆ Future advantages of the programme for the members of the family.
- ☆ Compatibility of the programme with their existing situation and customs.
- ☆ Cost and labour involved in relation to the return.

- ☆ Timeliness of the recommendations in relation to operation.
- ☆ Complexity of the programme and ease or difficulty with which the change can be made.
- ☆ Confidence of the people in the extension worker and the organization the represents.
- ☆ Approval or encouragement by opinion leaders and others who are held in high esteem.

Chapter 22
Rapport Building

Execution of the programme starts with environment building *i.e.* through rapport building and success of programme lies in involvement of local leaders, local bodies, organizations and extension agencies.

Rapport Building

Programme execution starts with Rapport Building with community as it is easier to get what you need from someone and to find out what they need, if one has a good rapport with someone. Rapport building is a process of establishing relationship, affinity or alliance with members of the social system and the extension administrator. It is an ongoing part of the intervention, not just a stage that occurs once at the beginning and is then stopped. Rapport is one of the most important features of subconscious communication. It means being "in sync" with, or being "on the same wavelength" as the person with whom you are talking.

The better rapport that you have in a relationship with someone, the more you can ask of them. Rapport building is not a mechanical set of procedures but is based on sharing, cooperation, and mutual, give-and-take. It is characterized by exchange of information enriched by creating credibility of competence, trustworthiness and empathy with members of the social system. Rapport is therefore an important part of a communication based approach to intervention, which assumes a speaker and a listener who want to interact with one another. It is through rapport building process that much of the adoption or rejection of technology depends upon by members of the social system. It is one of the major outcome of rapport building process that extension administrator is treated as a part of the social system or not.

Objectives of Rapport Building

- ☆ Identify oneself with members of the social system with trust and competence.
- ☆ Identify the project, objective of the programme and sharing same with the members of the social system to develop mutual understanding and concerned for the programme.
- ☆ Know the role, responsibility and partnership function with the programme.
- ☆ Develop trustworthiness and feeling of empathy from both sides *i.e.,* extension worker and rural people.

Steps for Rapport Building

- ☆ Before visiting village make list of issues to be discussed, list of persons to be met and list of institutions, which are working over there.

- ☆ Empathize with someone's emotional state by being on 'their side' (also called as mirroring). Take a genuine interest in getting to know what's important to the other person. Start to understand them rather than expecting them to understand you first. Be a good listener so as one can listen for key words and problems that arise when talking to the person. While doing so question them to better your understanding of what they are saying and show your empathy towards them.
- ☆ Match the tone of a person's body language in terms of your body language, gestures, voice tone and speed. Match the tone, tempo, inflection, and volume of a person's voice.
- ☆ Respect the other person's time, energy, favorite people and money. They will be important resources for them.
- ☆ Deliberately finding something in common with a person in order to build a sense of trust. This is done through shared interests, dislikes, and situations.
- ☆ Note how someone deal with information or likes to handle information *i.e.* do they like lots of details or just the big picture
- ☆ As you speak feed-back information in this same portion size.

Involvement of Local Leaders, Local Bodies, Organizations and Extension Agencies

Community power structure is the complex network of relationships between the recognized power holders and the interplay of their roles in a community. Thus success of any programme lies in participation and total involvement of the people in decision making process. It includes full participation of various village development unions, age grades, individuals, philanthropic organizations, local leaders, local bodies, institutions and extension agencies.

Chapter 23

Constraints in Programme Implementation

No matter how well the programmes have been developed, there lie certain constraints in implementation of programme at grass root level. Such as:

- ☆ Lack of qualified manpower with appropriate experience in agricultural extension: Village workers or extension worker lack appropriate professional qualification to meet the demands of the new and sophisticated agricultural technology that hinders in transfer of technology.
- ☆ Low extension coverage rate: Extension worker have larger area and larger number of families to work with which they sometimes are unable to reach or attend properly.
- ☆ Improper selection of Contact farmer: Contract farmer is important link between farmers and the

extension community. Thus success of programme depends on them. Several studies show that selection and use of contact farming is not complete. Mostly well to do farmers are selected thus farmers with small land holding lag behind.

- ☆ Lack of timely visit of village level workers and scientists to village for solving their problems.
- ☆ Lack of co-ordination with other agencies involved in agricultural production.
- ☆ Over-reliance on extension through mass media approach; field activities that motivate farmers to experience improved/innovative practices being kept to a minimum due to limited resources.
- ☆ Insufficient support to rural women intensively involved in farming and rural youth.
- ☆ Fragmented land holding leads the farmers to less invest in land and land improvement.

Ways to Overcome Problems in Programme Implementation

Decentralization of extension programme planning is the only solution to overcome programme implementation. It means transferring specific decision-making functions to local levels, authority to be transferred to rural people including the financial planning and involvement of local government.

- ☆ Strategies need to be built for development of landless labourer and farmers with small holdings. Due to population explosion gradual decrease in farm size is observed. Consequently the farmers will eventually less inclined to invest in land and land improvement which may lead to drop in productivity and lowered economic visibility of farms.

- ☆ Extension worker need to work cooperatively with rural credit agencies because besides providing financial assistance, these institutions can advise and assist extension educators on programme priorities, plan and implementation strategies.
- ☆ Holistic problem solving approach need to be followed by agricultural and extension institution so as to find solutions through multi disciplinary efforts.
- ☆ Linkage need to be established between professionals in the discipline of agriculture education, extension education sociology economics anthropology for both collaborative work and practical problem solving
- ☆ Active participation of rural people in planning, implementing and monitoring of extension programs need to be encouraged by extension worker which will change the nature and direction of development intervention.

Chapter 24

Monitoring and Evaluation of Extension Programme

Extension programmes are mostly funded with public money and are planned and implemented by an organization, which in most cases is a department of government. In order to justify the appropriation of public funds and continuing support from the people, it is necessary that their management as well as impact be properly and adequately monitored and evaluated from time to time.

Monitoring

The word monitoring is derived from Latin word "monitor" meaning to warn. Monitoring is the systematic collection and analysis of information as a project progresses. It represents an on-going activity to track project progress against planned tasks. It is aimed at improving the efficiency and effectiveness of a project or organisation.

It helps to keep the work on track by keeping regular oversight of the implementation of an activity in terms of input delivery, work schedules, targeted outputs, etc. through routine data gathering, analysis and reporting. It enables you to determine whether the resources available are sufficient and are being well used, whether the capacity is sufficient and appropriate, and whether work is progressing as planned.

Importance of Monitoring

- ☆ Provides project management *i.e.*, keeps check on whether progress is being made towards achieving project objectives.
- ☆ Provide regular feedback to enhance the ongoing learning experience and to improve the planning process and effectiveness of interventions.
- ☆ It alerts managers regarding actual and potential project weaknesses, problems and shortcomings before it is too late.
- ☆ It provides managers with the opportunity to make timely adjustments and corrective actions to improve the programme/project design, work plan and implementation strategies.

Evaluation

The word 'evaluation' has its origin in the Latin word *valere* meaning to be strong and valiant. Its dictionary meanings are the "determination of the value," the "strength" or "worth of something," and appraisal, an estimate of the force or making a judgment of something.

Evaluation is the systematic acquisition and assessment of information to provide useful feedback about some object. It also refers to a periodic process of gathering data and then analyzing it in such a way that the resulting information can be used to determine whether the program is effectively carrying out planned activities. An evaluation can also illustrate the extent to which the program is achieving its stated objectives and anticipated results.

Definitions of Evaluation

Here are few more specific definitions of evaluation as given by scientists involved in rural development programmes.

- ☆ It is a process of systematically drawing upon experience as a means of making future efforts more effective (Beaglehold, 1955).
- ☆ Evaluation is a coordinated process carried on by the total system and its individual subsystem. It consists of making judgments about a planned programme based on established criteria known and observable evidence (Boone, 1985).
- ☆ Programme evaluation is the determination of the extent to which the desired objectives have been attained or the amount of movement that has been made in the desired direction (Boyle and Johns, 1970).
- ☆ Evaluation is a comparison of the situation before and after a development programme has operated within it for a predetermined period (Matthew, 1956).
- ☆ Programme evaluation is the process of judging the worth or value of a programme. The judgment is formed by comparing evidences as to what the

programme is with criteria as to what the programme should be (Steele, 1970).

- ☆ The process of determining the extent to which objectives have been attained is evaluation (Thiede, 1964).

Importance of Evaluation

- ☆ Provide managers with information regarding project performance
- ☆ Evaluation provides signs of project strengths and weaknesses enabling the managers to improve future planning and decision-making.
- ☆ Evaluations can verify if the program is running as originally planned as project plans might change during the implementation process.
- ☆ Assist in determining the degree to which service objectives are accomplished.
- ☆ Evaluation during mid-term may serve as a means of validating the results of initial assessments obtained from project monitoring activities.
- ☆ Evaluation conducted after the termination of a programme/project determines the extent to which the interventions are successful in terms of their impact and sustainability of results.
- ☆ Generate detailed information about project implementation process and results for public relations, fundraising, promotion of services in the community, as well as identifying possibilities for project replication.
- ☆ Improves the learning process as evaluations often explain the causes as to why activities succeeded or failed.

- Justifies and shows accountability of services.
- Provides input for planning and development of future services relevant and effective.

Types of Evaluation

There are many different types of evaluations depending on the object being evaluated and the purpose of the evaluation. Perhaps the most important basic distinction in evaluation types is as follows:

- *Ex-ante evaluation*: Evaluation carried out before the project begins. It helps to prepare base for development of the project.
- *Formative Evaluation or "Interim" evaluation (mid-term)*: This type of evaluation monitors a service, program, or resource and is intended to assist decision makers improve that service while it is in operation.
- *Summative Evaluation*: This type of evaluation occurs at the completion of a project, service, or program and is intended to prove the quality, worth, or impact from that effort.
- *Ex post evaluation*: This type of evaluation occurs after completion of the programme/project

Classification of Evaluation Based on People Involved in it:

- *Internal evaluation*: Internal evaluation is a process of quality review undertaken within an institution for its own ends, with or without the involvement of external agents. In internal evaluation, the programme or project members analyse their own work from the internal standpoint.

☆ *External evaluation*: The word "external" means that the evaluators are not project members. It is a process that uses people external to the programme or institution to evaluate quality or standards. "External evaluations are therefore suitable for factual issues, relatively complex situations, and overall appraisals.

Difference between Internal and External Evaluation

Internal Evaluation	*External Evaluation*
Presents internal view	Provide external view
Evaluation object lies within ownsphere of responsibility	Evaluation object lies within other's sphere of responsibility
The evaluators are very familiar with the work, the organisational culture and the aims and objectives.	Incorporation of knowledge of similar actions and themes
It is periodic process that is possible at regular and shorter intervals	It has limited time frame and is specific, often one-off action
There is proximity to subject and to the project	Distance from subject/project
It is more economical	More cost-/effort-intensive

Methods of Evaluation

Document Analysis

It includes analysis of minutes of meetings, correspondence, budget records, workshop notes, participant papers, and newspaper reports etc. The data is analyzed for content, and summarized in relation to questions, including extent of inputs into the programme; levels of participation, nature of goals and activities, and themes regarding problems, concerns, expectations, and new directions. Themes from documents can be a credible source of information.

Observation Method

Both qualitative and quantitative data can be collected through observation method. Observers can be outsiders or persons who are involved in programme activities. This method evaluates best the extent of participation and personal interaction, nonverbal indicators of interest or inattention, leadership roles, performance levels, and conflict indicators. These data are very powerful graphic ways of communicating the nature of a programme and its outcomes to sponsors as well as to potential learners.

Interview Method

Interviews are one of the most widely used methods for programme evaluation. Interviews with key informants are suited to in-depth explorations of issues. Responses of standardized and closed-ended questions can be tabulated numerically to indicate item strength. Responses of open-ended questions which provide information regarding reasons why the activities are viewed differently by diverse groups of participants. Illiterate members of the community can participate fully through interviews.

Survey Method

The survey method of evaluation is a more standardized form of data collection that incorporates a prepared questionnaire. Surveys are often used to evaluate the extent of practice, estimations of production yields, preferences for appropriate technology, and expectations regarding the future. They are best used with homogeneous populations rather than with quite diverse populations because standardized questionnaires are less likely to be sensitive to diversity. Evaluation of unique practices and adaptations of diverse farmer types is best done through interviews and observations.

Field Visit Method

Field visits and tours are the best method to provide authenticity and reality to the conditions, limitations, and impacts of extension programmes. During field visit evaluators comprise of the local farmers, extension staff, administrators, funders, and external evaluators that provide balance and interactive learning regarding different perspectives. Assignments are distributed so that special knowledge regarding specific aspects of situations can be gathered. Some team members focus on economic, social, and cultural aspects, while others focus on technical aspects. Comparing data, analysis and reflection on findings, and insights following field visits generate a more holistic, balanced evaluation.

Village Drama and Role Plays

Performance of villagers in drama or role play describes the interaction process of extension staff with the village on a specific practice that reveal a variety of evaluative data on social relationships, relevance of extension knowledge to local knowledge, and historical events that have affected the solutions to farmer problems.

Case Studies

Case studies of specific types of farmers or farming practices, helps to understand motivation of farmers or potential extension contacts. Comparisons between farmers who have used extension technologies and those who have not are common types of case studies. The typology may be based on geographic regions, soil types, and cultural, age, gender, and economic differences. Case studies are best constructed through repeated interviews over time and often include, in addition to self-report, data from persons who know the subjects well. Oral histories, logs, and journals can also contribute to case study data if farmers collaborate in producing these case studies. Evaluators should guarantee the right to privacy and confidentiality of their sources. Case studies often reveal deterrents to participation, as well as ways participants have overcome deterrents to practice through extension contacts.

Chapter 25
Training of Leaders

A programme can be said to be successful only when it prepares few villagers to lead the programme even after the extension workers have left. This can be done by identifying right leader for leadership by extension agent. Thus it is important for extension workers to know about types of leaders, their quality and role.

Leader

A leader is "a person who guides and directs others. He has capability to inspire and influence a group of people *i.e.,* followers towards the achievement of a goal". A leader influences a group of people towards a specific result. It is not dependent on title or formal authority.

Leadership

Leadership is characteristic of leader. It is a force that binds a group together and motivates it towards the desired goals. Leadership is about being successful themselves and enabling success in others *i.e.,* followers.

Qualities of a Good Leader

- ☆ *Intelligence*: A good leader is cleaver enough to solve problems of the group.
- ☆ *Dominance*: A leader has governance power *i.e.,* has power to control the group decision in a desired direction.
- ☆ *Adaptability*: A good leader is flexible in his decisions looking at the need/interest of the group; he changes his plans and ideas.
- ☆ *Persistence*: Firm determination to fight or work for the benefit of the group is one of the major characters of a good leader.
- ☆ *Integrity*: Honesty is the major policy of the leader; therefore the members rely fully on their leader's decision.
- ☆ *Self-confidence*: A group leader possesses confidence to assure success to the group through his decisions.
- ☆ *Conscientiousness*: A good leader pays attention to every aspect of the problem and the alternative solutions to it. He goes through every fine details to overcome the problem.
- ☆ *Openness to experience*: A good leader is confident to try for new inventions and ideas. He has risk taking ability.
- ☆ *Risk bearing capacity*: Ability to share risk or hardship on an equal foot.

- *Respectable*: A leader wins respect without courting popularity
- *Leader lead by examples*: A good leader practices what he preaches.
- *Good listener*: A good leader is always willing to discuss and solve the problem. He gives time to listen with understanding to the suggestions and alternative solutions, given by the group members, to a problem.

Role of Leader

- To inspire and create followers who are also self-leaders through one's ability to negotiate and communicate, influence and persuade others to do things in the same manner.
- To forms vision *i.e.*, he knows what need to be achieved and how thus he provides people with bridge to the future.
- To build trust that makes an organization to work effectively. He speaks honestly and openly and expects the same from others. He has faith in suggestions of others and thus communicates openly and honestly to them. He relies on judgment of the group and vice-versa.
- To combine personal competence with competencies of others by having common touch with the group members and understanding each individual's abilities, limitations, motives etc. This helps in bringing best out of the group.
- To keep track of the progress against the target and managing the group action.

Identification of Type of Leader

One of the major factors determining success of any programme depends upon identification of leader in the initial stage. With selection of wrong leader, it becomes impossible afterwards to build a winning team. Thus one needs to know the type of leaders and their identification according to their specific characteristics and their competence in varied situation.

1. Authoritarian or Autocratic Leader

Under this leadership style, all decision-making powers are centralized in the leader.

It is characterized by:

- Leaders making the decisions alone.
- He demands strict compliance to his orders.
- He decides each step to be taken
- He does not trust people
- His every step is uncertain to a large degree.
- They do not entertain any suggestions or initiatives from subordinates.
- It permits quick decision-making, as only one person decides for the whole group and keeps each decision to himself until he feels it is needed to be shared with the rest of the group.
- The leader is not necessarily hostile but is aloof from participation in work and commonly offers personal praise and criticism for the work done.
- The authoritarian leadership style, for example, is approved in periods of crisis but fails to win the "hearts and minds" of their followers in the day-to-day management.

- Example: In an emergency when there is little time to converge on an agreement and where a designated authority has significantly more experience or expertise than the rest of the team, an autocratic leadership style may be most effective.

2. Democratic Leader

The democratic leadership style favours decision-making by the group and in it leader gives instruction after consulting the group.

It is characterized by:

- Collective decision processes, assisted by the leader.
- Before accomplishing tasks, perspectives are gained from group discussion and technical advice from a leader.
- Members are given choice and collectively decide the division of labour.
- Praise, encouragement and criticism are the objective.
- The democratic leadership style is more adequate in situations that require consensus building.
- They can win the co-operation of their group and can motivate them effectively and positively.
- The decisions of the democratic leader are not unilateral as with the autocrat because they arise from consultation with the group members and participation by them.

3. Laissez-Faire Leader

Here leader does not lead, but leaves the group entirely to itself. They allow maximum freedom to subordinates, *i.e.*, they are given a free hand in deciding their own policies and methods.

It is characterised by:

- ☆ Freedom given to the group for policy determination without any participation from the leader.
- ☆ The leader remains uninvolved in work decisions unless asked *i.e.*, the leader does not "take charge".
- ☆ Leader does not participate in the division of labour.
- ☆ Here leader is very infrequently gives praise.
- ☆ He may withdraw from responsibility anytime.
- ☆ He can be perceived as a failure in protracted or thorny organizational problems.
- ☆ Example: in a highly motivated and aligned team with a homogeneous level of expertise, a more democratic or laissez-faire style may be more effective.

4. Toxic Leader

Toxic leader is one who has responsibility over a group of people or an organization, and who abuses the leader and follower relationship by leaving the group or organization in a worse-off condition than when s/he first found them.

5. Professional Leader

Professional leader is one who is trained for taking leadership positions for management in institutions, organizations etc. They have undergone specific training/ certificate courses. For example, in the case of a school, the principal is trained leader for the management of studies in school. Similarly in kindergarten the pre-school-teacher is the trained leader to manage the play or teachings for the pre-schoolers.

6. Lay Leader

Lay leader is one who may/may not have undergone professional training. They are mostly the local leaders who volunteer for responsibility. They are paid for their work.

Training

After identifying the leader, they can be trained through various methods for carrying forward the programme. The term training is derived from word "train" that means 'practicing different learning again and again'. An effective training starts from creating a feeling of need and interest among its members.

There are various training method depending upon the nature of group and objective of the training. The three basic training methods are:

1. *Lecture method*: It is an oral/printed presentation in which the instructor or other speakers present facts, opinions, events, principles, or explanations. It is used most often for giving orientation, directions or to introduce material, explanation of basic concepts and for illustration and review. This method is used to create understanding of a topic and influences behavior, attitudes of the trainees. It is oldest, easiest, versatile and least expensive method. It is traditional form of training. It is least efficient method as effectiveness of lecture lies in passive nature of the trainees. It still dominates the training programme.

2. *Demonstration method*: The demonstration may be live or pre-recorded. This method is a visual display of how something works or how to do something. The learners observe the performance of a task, operation, or procedure. It helps the trainees to focus their attention on critical aspects of the task It is often followed by performance or practice on the part of the learner

3. *Participatory training method*: This method gives practical training to the trainees. The various methods in this allow the trainee to participate in a real fashion. It is about imitating or making judgment or opining how events might occur in a real situation. It is best used for skill development. Various games and stimulation giving activities are planned in which trainees compete with each other according to the defined set of rules. It includes business games, role play etc.

The five elements of training are: Teacher, Learner, Subject matter, Teaching aids and physical facility. Combination of these elements makes an effective training.

By using greater number of sensory channels, training enhances knowledge, skill, attitude, experience and behaviour of the trainees.

1. Knowledge increases up-to-date information and self-confidence.
2. Skill helps in better job performance by increasing self- efficiency
3. Attitude helps to built rapport.

4. Experience increases better competence
5. Behaviour brings greater conformity and better promotional perspective.

Chapter 26

Participatory Rural Appraisal: Concept and Techniques

In 1983, Robert Chambers, a Fellow at the Institute of Development Studies (UK), used the term Rapid Rural Appraisal to describe techniques that could bring about a 'reversal of learning' *i.e.* U-turn or turnaround learning. By the mid 1990's, the term RRA had been replaced by a number of other terms including 'Participatory Rural Appraisal (PRA)' and 'Participatory Learning and Action' (PLA).

The difference is that PRA emphasizes processes which empower local people, whereas RRA is mainly seen as a means for outsiders to gather information. According to Robert Chambers "Participatory Rural Appraisal is a family of approaches and methods to enable local people to share, enhance, and analyse their knowledge of life and conditions, to plan, and to act."

It is an approach to include the knowledge and opinions of rural people in the planning and management of rural development projects and programmes. To ensure that illiterate rural people are not excluded from participation, these techniques avoid writing instead rely on the tools of oral communication like pictures, drawing symbols and group memory.

The process of PRA consists of wide range of tools such as: Individual, Group interview and direct observation.

Principles of PRA

- ☆ Using optimal ignorance- means to avoid or ignore unnecessary and irrelevant data and selecting the relevant one.
- ☆ Off-setting bias- means listening to all irrespective of caste, sex, economic status etc. It means to avoid biasness.
- ☆ Triangulation- means using more than one technique/source of information to cross-check answers. This is done by comparing and complementing information from different sources gathered in different ways.
- ☆ Learning from and with the people. It means learning directly from the rural people through their participation in information generation.
- ☆ Optimizing trade-offs: Trade-offs means not merely obtaining accurate mathematical ratios, but there should be cost-effectiveness, alertness, observation, imagination and the ability to pursue serendipity.

PRA Techniques and Methods

The method of PRA is group based intensive activity. For conducting PRA the extension worker should be good listener. The most common methods are as follows:

1. Mapping

It is a visual representation of what the community perceives as their community space. It shows the shape (appearance) of the community, boundary and all the major features as understood and known by the community. It is used for understanding of geographical features. It shows location of buildings, dwellings based on caste, social relations etc.

There are different maps for different purpose. For instance:

a. *Social map*: It is map representing households according to certain indicators.
   - ☆ Indicates where people live and how many people live in an area
   - ☆ Social and residential differences in status and wealth
   - ☆ Buildings where people live or work, uses of space in a house

b. *Resource map*: It shows location and use of natural resource in an area. Example: location of water, field etc.

c. *Physical map*: It shows location of school, roads, health facilities etc.

2. Time Line

It is used to trace important events of past. It is a list of key events in the history of community. It provides information on various land marks in development of village for example: institution development, educational and infrastructural development etc. Elderly people are mainly asked for collection of data. The time line helps the team to understand what local or national events the community considers to be important in its history.

3. Transact Walk

It is a walk or series of walks through an area with local informants. It shows a cross section of the area as observed by the walk thus it is also called cross-sectional walk. It helps to gain firsthand knowledge and to improve upon various maps. It adds detail on specific characteristics (slope drainage, vegetation, water, soils other sources) that further verifies the PRA team's understanding of the area.

4. Resource Inventory

It is used to collect information about all resources present in the village such as human resource, animal and material resources etc.

5. Venn Diagram

It is also called chapatti diagram. It helps to find out functioning and usefulness of institutions. Its purpose is to identify group and institutions operating in community and to show how they interact with each other. It also shows the degree of their cooperation and involvement in development programs and their influence on decision making in the community.

Importance of PRA

- ☆ Provides basic information in situations where little is known

- ☆ Identifies and assess problems
- ☆ Helps in appraising, designing, implementing, monitoring, and evaluation programs and projects
- ☆ Gives a better picture of needs and organizations' ability to meet village needs
- ☆ Helps in developing and transferring appropriate technologies
- ☆ Helps in planning of projects that are more relevant, restructuring administrations, assisting in decision-making and policy formation
- ☆ Provides guidelines for survey designs and assess the applicability of their results to other places.

Chapter 27

Concept of Five Year Plans and Rural Development

For the smooth functioning of any economy, planning plays an important role. India planning is undertaken by the Planning Commission of India in order to improve the standard of living of the people belonging to the country. To improve the quality of life of the people, the resources of the country have to be utilized efficiently.

The Planning Commission was set up in March, 1950 by a Resolution of the Government of India. The economy of India is based in part on planning through its five year plan, developed, executed and monitored by the Planning Commission. With the Prime Minister as the ex-officio Chairman, the commission has a nominated Deputy Chairman, who has rank of a Cabinet minister. The tenth plan ended its tenure in March 2007 and the eleventh plan is currently underway.

For rural planning, India planning has set up a Rural Planning and Credit Department which is under the Reserve Bank of India. This department formulates polices which help in rural development. It also extends credit for employment programs and agricultural activities in rural areas. Under India planning, local bodies have been set up in all states in order to develop urban areas. These urban local bodies such as the municipal corporation and the Panchayat see to it that the services and basic infrastructure are well maintained.

Five Year Plan	*Duration*	*Five Year Plan*	*Duration*
First	1951-1956	Seventh	1985-1990
Second	1956-1961	Eighth	1989-1991
Third	1961-1966	Ninth	1992-1997
Fourth	1969-1974	Tenth	1997-2002
Fifth	1974-1979	Eleventh	2007-2012
Sixth	1980-1985		

The first five year plan of India (1951-1956) had been presented by the then Prime Minister Jawaharlal Nehru in the Indian Parliament on 8th December, 1951. It had been made by the planning commission whose objective was to improve the standard of living of the people by effective use of the country's resources.

The various tasks of the second five year plan (1956-1961) in India were to increase by 25 per cent the national income, to make the country more industrialized and to increase employment opportunities so that every citizen gets a job.

The aim of third five year plan (1961-1966) was to increase the quality of life of the citizens through effective use of the country's resources.

During Fourth Five-Year Plan, (1969–1974) Indira Gandhi was the Prime Minister. The Indira Gandhi government nationalized 14 major Indian banks and the Green Revolution in India advanced agriculture. In addition, the Indo-Pakistan (now Bangladesh) War of 1971 and Bangladesh Liberation War took place. Funds earmarked for the industrial development had to be diverted for the war effort.

At the onset of the Fifth Five Year Plan (1974-1979) India in the 1970s, the international economy was in a turmoil, which had a great impact on the economy of both, developed and developing countries of the world. The Fifth Five Year Plan India was designed with emphasis on objectives, such as: to reduce social, regional, and economic disparities for developmental planning, to enhance agricultural productivity, to initiate land reforms etc.

The objectives of the Sixth Five Year Plan (1980-1985) India were mainly focused on increasing industrialization and reducing long-standing problems such as poverty and unemployment.

The Seventh Five Year Plan India was for the duration between 1985 and 1989 under the approval of the National Development Council in India. As an outcome of the sixth five year plan, there had been steady growth in agriculture, control on rate of Inflation, and favorable balance of payments which had provided a strong base for the seventh five year plan to build on the need for further economic growth. The 7th Plan had strived towards socialism and energy production at large.

Just before the formulation of the Eighth Five Year Plan *i.e.* (1992-1997), there was great political instability in India which hindered the implementation of any five years plan for the following two years after the Seventh Five Year Plan. This period is characterized by extreme FOREX reserve crisis and introduction of liberalization and privatization in Indian economy.

The Objectives of Ninth Five Year Plan India (1997-2002) was: to prioritize agricultural sector and emphasize on the rural development, to generate adequate employment opportunities and promote poverty reduction, to stabilize the prices in order to accelerate the growth rate of the economy, to ensure food and nutritional security etc.

The Tenth Five Year Plan India (2002-2007) aims to transform the country into the fastest growing economy of the world and targets an annual economic growth of 10 per cent. This was decided after India registered a 7 per cent GDP growth consistently over the last decade.

The Eleventh Five Year Plan India (2007-2012) has the following objectives:

1. Income and Poverty

- ☆ Accelerate GDP growth from 8 per cent to 10 per cent and then maintain at 10 per cent in the 12th Plan in order to double per capita income by 2016-17
- ☆ Increase agricultural GDP growth rate to 4 per cent per year to ensure a broader spread of benefits
- ☆ Create 70 million new work opportunities.
- ☆ Reduce educated unemployment to below 5 per cent.
- ☆ Raise real wage rate of unskilled workers by 20 percent.

- ☆ Reduce the headcount ratio of consumption poverty by 10 percentage points.

2. Education

- ☆ Reduce dropout rates of children from elementary school from 52.2 per cent in 2003-04 to 20 per cent by 2011-12
- ☆ Develop minimum standards of educational attainment in elementary school, and by regular testing monitor effectiveness of education to ensure quality
- ☆ Increase literacy rate for persons of age 7 years or above to 85 per cent
- ☆ Lower gender gap in literacy to 10 percentage point
- ☆ Increase the percentage of each cohort going to higher education from the present 10 per cent to 15 per cent by the end of the plan

3. Health

- ☆ Reduce infant mortality rate to 28 and maternal mortality ratio to 1 per 1000 live births
- ☆ Reduce Total Fertility Rate to 2.1
- ☆ Provide clean drinking water for all by 2009 and ensure that there are no slip-backs
- ☆ Reduce malnutrition among children of age group 0-3 to half its present level
- ☆ Reduce anemia among women and girls by 50 per cent by the end of the plan

4. Women and Children

- ☆ Raise the sex ratio for age group 0-6 to 935 by 2011-12 and to 950 by 2016-17

- ☆ Ensure that at least 33 per cent of the direct and indirect beneficiaries of all government schemes are women and girl children
- ☆ Ensure that all children enjoy a safe childhood, without any compulsion to work

5. Infrastructure

- ☆ Ensure electricity connection to all villages and BPL households by 2009 and round-the-clock power.
- ☆ Ensure all-weather road connection to all habitation with population 1000 and above (500 in hilly and tribal areas) by 2009, and ensure coverage of all significant habitation by 2015
- ☆ Connect every village by telephone by November 2007 and provide broadband connectivity to all villages by 2012
- ☆ Provide homestead sites to all by 2012 and step up the pace of house construction for rural poor to cover all the poor by 2016-17

6. Environment

- ☆ Increase forest and tree cover by 5 percentage points.
- ☆ Attain WHO standards of air quality in all major cities by 2011-12.
- ☆ Treat all urban waste water by 2011-12 to clean river waters.
- ☆ Increase energy efficiency by 20 percentage points by 2016-17.

References

Http://www.managementparadise.com/forums/articles/13051-basic-techniques-building-rapport.html

Http://www.wcd.nic.in/icds.htm

Http://nird.ap.nic.in/clic/list.html

Http://icarzcu3.gov.in/atic.htm

Http://scstobc.org/csindiraimy.html

Http://www.sociologyguide.com/women-and-society/rashtriya-mahila.php

Http://rmk.nic.in/chap1.htm

Http://www.managementstudyguide.com/importance-of-communication.htm

Http://www.andrews.edu/~tidwell/bsad450/communicationbarriers.html

Http://business.mapsofindia.com/india-planning/

Http://www.krishiworld.com/html/agri_extension_edu2.html

Http://www.extension.iastate.edu/Publications/EDC118.pdf

Http://www.ces-vol.org.uk/index.cfm?Pg=543

Http://www.vtaide.com/gleanings/leader.htm

Http://www.businessballs.com/leadership.htmc

Http://www.icarzcu3.gov.in/kvk/technical/5.pdf

Http://www.pdfcari.com/OVERCOMING-BARRIERS-TO-COMMUNICATION.

Extension Education and Community Development. Directorate of Extension, Ministry of Food and Agriculture , Government of India, New Delhi, 1961.

Dhama. O.P. Extension and Rural Welfare 5th Edition, Ram Prasad and Sons, Agra, 1968

Kelsey, L. D. and Heame, C.c. Cooperative Extension Work.comstockpublishing Associates, Ithaca, New York, 1949

Wilson and Gallup, Extension Teaching Methods, U.S.A. Department of Agriculture Washington D.C. 1949.

Evolution of Community Development Programme in India. Ministry of Community Development , Cooperation and Panchayati Raj.

Moore, Social Change.Prentice Hall of India Pvt., New Delhi

Sharma, R.N., Principles of Sociology. Asia Publishing House, New Delhi, 1968.

Devdas, Rajammal P., Meaning of Home Science, Coimbatore, 1958.

Brown, Lewis and Harcleored, A.V. Instruction.

Leagans J. Paul. Criterion for Procedures and Practices in County Extension Programme Development, Cornell University, New York, 1950.

R.Dayal, Community Development Programmes in India.

Evaluation Reports on Community Development and National Extension Service, Programme, Evaluation, Organisation Planning Commission, Government of India.

Hatcher, H.M., and Andrew, The Teaching of Home Making.

Saville, A.H. Extension in Rural Communities, Oxford University Press, 1965.

Radhamurthi V., Extension in planned Social Change: Allied Publishers, Madras, 1964.

Chiovoloni M 2004. The interactive power of local and traditional communication systems. *Ileia Magazine.* April 28, 2004, pp. 1-4.

(Arnold, E and Boggs, K. 2007). Rapport building steps. (Mirroring)

Rural socio-economic change in India-Implication for agriculture extension; Rama Radhakrishnan; Research Asoociate; The Pennsylvania State University; Http://www.fao.org/docrep/w5830e/w5830e0j.htm

Overcoming barriers to communication by Bernard L. Erven. Department of Agricultural, Environmental, and Development Economics. Ohio State University Html

Extension Communication and Management by Ray, G.L.

Extension Education by Supe, S.V.

What is Home Science – Kalpana Parlikar

Training Manual on Participatory Rural Appraisal Compiled by Simon Adebo

(Freelance Consultant) December 2000, Addis Ababa.

Sample Question Papers

Home Science Extension & Rural Development

Ques. 1: Write Short Note on

1. Role of Extension in Rural Development
2. Role of ICAR in Rural Development
3. Objective of Home Science Extension
4. Principle of Participation
5. Principle of cultural difference

Ques. 2: Briefly ANY TWO of the following:

1. NAIP *OR* IVLP
2. ATIC *OR* SJGSY

Ques. 3: Explain in details ANY ONE of the following

Panchayati Raj Institution OR

Community Development Programme

Ques. 4: Fill in the blanks with WORDS GIVEN in brackets:

1. Panchayati Raj Institution is _______ tier system. (*Three/Five*)
2. _______ of Home Science Extension is "helping people to help themselves". (*Principle/Philosophy*)
3. Shriniketan Project was started by _______ in Bengal in the year 1921. (*Rabindra Nath Tagore/F L Bryne*)
4. Extension education is _______ education. (*Formal/Non-formal*)
5. Nilokhedi Experiment was done by Shri S K Dey at Krukshetra that aimed at _______. (*rehabilitation of people during partition/prohibition of untouchability*)
6. According to Principle of _______, extension programme should start from local situation, local group and local problems. (*Cultural Difference/Grass Root*)
7. The objective of _______ Programme is to prevent, reduce and eradicate morbidity and malnutrition among vulnerable mother and children. (*TRYSEM/ICDS*)
8. _______ programme is for those women and girls who have no house or have family/social problem or mental tension etc. (*Short Stay Home/Vande Mataram Yojana*)
9. _______ Education is life long process by which every person acquires and accumulates knowledge

and skill from daily experience. (*Informal/Non-formal*)

10. Indian Council of Agricultural Research (ICAR) was established on 16 July _______ (*1929/1959*).

Ques. 5: Write True or False for the statements given below:

1. There are eleven principles for Home Science Extension
2. The term 'Extension' originated in England, 1866.
3. Panchayati Raj Institution was introduced on recommendation of Balwant Raj Mehta Committee in the year 1959.
4. Under Applied Nutrition Programme (ANP) vitamin A & Anaemia prophylaxis programme was started and supplementary nutrition for infant and Toddlers & Pre schooler's were started.
5. Extension is defined as "science of developing capability of people for sustainable improvement in their quality of life".
6. Rural Development is dynamic process
7. Community Development Programme was launched in India on 2nd October 1952.
8. The Beneficiaries of SJGSY programme are called "Swarojgaries"
9. The objective of Sampoorna Gramin Rojgar Yojana (SGRY) is to provide additional wage employment in rural areas.
10. Home Science Extension is multi-dimensional and inter-disciplinary in nature.

Ques. 6: Match the Column A with Column B

Column A	Column B
1. Extension	a) School & College Education
2. Principle of Indigenous Knowledge	b) Mr. Albert Mayer
3. Rural Development	c) Sarpanch
4. Formal Education	d) Beneficiaries are 18-35 year old Rural youth
5. Etawah Pilot Project	e) To change attitude of society towards girl child
6. Village Panchayat	f) Promote and assist voluntary action in implementation of project for enhancing rural prosperity.
7. TRYSEM	g) Non-Government Organisation
8. Balika Samriddhi Yojana	h) Derived from Latin word 'ex' and 'tensio' that means out stretching
9. CAPART	i) Extension agent should understand traditional knowledge which people have developed through generations of work experience
10. NGO	j) Complete transformation of rural life for betterment

Ques. 7: Write expansion (Full Forms)of following Abbreviations:

1. HYVP
2. ATMA
3. NAIP
4. ICDS
5. IVLP

Communication and Diffusion of Innovation

Ques. 1: Define the following:

1. Adoption
2. Noise
3. Diffusion
4. Rate of Adoption
5. Homestead Technology
6. Diffusion effect

Ques. 2: Write difference between:

1. Interpersonal Communication and Mass Media
2. Cosmopolite and Localite channel
3. Manifest consequences and Latent consequences
4. Desirable Effect and Undesirable Effect
5. Innovator and Laggards

Ques. 3: Fill in the blanks

1. ________ is process in which people continues to adopt an innovation vigorously when expert thinks that it should not be done.
2. ________ channels are more useful during awareness stage of innovation decision.

3. It is easy to communicate between people who are ________ in nature.
4. ________ is degree to which an innovation is changed or modified by a user in the process of its adoption and implementation
5. ________ is decision to reject an idea in order to adopt better innovation
6. Favourable and unfavourable attitude towards an innovation is formed during ________ stage of innovation decision
7. When an individual is exposed to an idea for the first time, then he is at ________ stage of innovation decision
8. At ________ stage of innovation decision, individual makes mental application of new idea
9. ________ are established behaviour pattern for the members of social system.
10. ________ are individual who are able to influence other individuals' attitude informally in desired way with relative frequency.

Ques. 4: Match the following:

Column A	Column B
1. Hardware technology	a) Respectable
2. Change Agent	b) Day Dreaming
3. Early Adopter	c) Skeptical/Doubtful
4. Internal Noise	d) Computer
5. Late Majority	e) Professionals

Ques. 5: What is innovation? Write characteristics of innovation which effect rate of adoption.

Ques. 6: What is felt need? What are the ways to identify these needs?

Ques. 7: What role does different communication channels play in diffusion of homestead technology? What are the problems faced in diffusion of homestead technology?

Instructional Technology

Ques. 1: Fill in the blanks

1. The term Technology is derived from two Greek words ________ and ________ that means 'skill' and 'science of study' respectively.
2. Medium group consists of ________ number of members.
3. Teaching is process that brings change in individual's ________
4. Visual Aid is an instructional device that can only be ________ but cannot be heard.
5. The first step of extension teaching is to capture people's ________
6. ABC of Poster stands for: ________, ________ and ________
7. The ultimate aim of extension teaching is to attain ________ by the learner.

8. Single sheet of paper folded twice or thrice is called ________
9. There are ________ elements of Teaching-Learning Process
10. ________ is visual aid that contains sandpaper on its back.
11. According to USE, Communication Methods can be classified as ________, ________ and ________.
12. __________ is thin sheet of transparent flexible material that can be projected by using OHP.
13. The central element of learning situation is ________.
14. Expansion of OHP is ________
15. E-mail stands for: ________

Ques. 2: Write True OR False for following statements

1. The main limitation of teaching through printed material is that illiterate cannot be taught
2. Bulletin Board is also called Tack Board
3. The major advantage of filmstrips is that it cannot be corrected easily
4. First hand information can be gained through individual contact method
5. Object is real thing which have been removed as unit from natural setting
6. Small group have member counting 15-25
7. Folder is single sheet of paper printed on both sides and generally distributed free of cost

8. One of the drawbacks of Mass contact method is that Rapport building is not possible
9. Slide projector is the best known still projector
10. Newsletter is mini-news paper of good quality.

Ques. 3: Match the Column A with Column B

Column A	Column B
1. Rapport can be built through	a) First time conducted in India in 1965 by Ministry of Food and Agriculture.
2. Quick and Large Coverage is possible through	b) Teacher, Learner, Physical facility, Subject matter, Teaching Aid
3. National Demonstration	c) Attention, Interest, Desire, Conviction, Action, Satisfaction
4. Example of Folk media	d) Individual Contact Method
5. Another name for Slide Projector	e) Mass Contact Method
6. Model	f) Diascope/Magic Lantern
7. Flash cards	g) Three-dimensional representation of real things
8. Pie Chart	h) Puppet Show
9. Elements of Teaching Learning Process	i) A set of small compact cards with illustration (Picture/photo) on the front side and useful information written on backside

10.	Steps of Extension Teaching Process	j)	Circle graph that is used to show how several parts make up whole

Ques. 4: Briefly explain ANY FOUR of the following

1. National Demonstration
2. Advantage of Audiovisual aids in Extension
3. Importance of Folk media in Extension
4. Use of Black Board in Extension
5. Use of Radio in Extension
6. Advantage of Individual Contact Method
7. Advantage of Mass Contact Method

Ques. 5: Distinguish Between ANY TWO of the following With Example:

1. Method Demonstration and Result Demonstration
2. Object and Model
3. Tree Chart and Flow Chart
4. Chart and Poster

Ques. 6: Explain in details ANY TWO of the following

1. Steps in Extension Teaching Process
2. Elements of Teaching- Learning Process
3. Factors affecting selection of extension teaching method
4. Group Contact Method (Advantages and disadvantages)

Programme Planning Implementation and Evaluation

Ques. 1: Fill in the blanks

1. _______ is a person who guides and directs others towards achievement of goals
2. During emergency _______ type of leadership style is most effective
3. Collective decision making is the characteristic of _______ type of leader
4. _______ Type of leader leaves the group in worse of its condition than when he first found it.
5. Principal of any intuition is example of _______ type of leader
6. Training is derived from the word train that means _______
7. An effective training starts from creating a feeling _______ and _______ among its members.
8. _______ training method is best used for skill development
9. _______ is outline of activities so arranged as to enable efficient execution of the programme.
10. _______ is process of consciously choosing course of action from available alternatives
11. _______ is degree to which an individual is able to influence other individual
12. There are _______ steps in programme planning.
13. Extension programme planning should have achievable _______
14. The first step in Extension Programme

Ques. 2: Write True or False for the following statements

1. Plan of work also include calendar of activity
2. Complexities of programme do not affect participation of the people.
3. Programme Planning requires advance thinking
4. Ex post facto evaluation helps to prepare base for development of the project
5. Decisions of autocratic leader are unilateral
6. Leader lead by examples
7. In homogenous group laissez fair leadership style is more effective
8. Professional leader are trained leaders from management institution and possess degree for it
9. Execution of programme begins with rapport building
10. One of the method to overcome problems of programme implementation is holistic problem solving approach.

Ques. 3: Match the Column A with Column B

Column A	Column B
1. Change Agent	a) It is idea/object that stimulates activity towards its attainment.
2. Opinion Leader	b) Expression of ends towards which our efforts are directed
3. Interest	c) It is the last of programme planning
4. Reconsideration	d) Person to whom people contact for his view regarding any problem

5. SWOT	e) Need
6. Objective	f) Predetermined course of action
7. Plan	g) Timeliness
8. Democratic Leader	h) Strength, Weakness, Opportunity, Threat
9. Factor affecting participation of people	i) Praise, criticism and encouragement are its characteristic
10. Difference between What is and What ought to be	j) Influences clients decision in direction desired by change agency

Ques. 4: Answer ANY TWO of the following in one sentence only

1. Extension Programme Planning has permanence with flexibility.
2. Extension Programme Planning has balance with emphasis.
3. Extension Programme Planning should select problems based on interest and need

Ques. 5: Explain ANY THREE of the following

1. Reason why Extension Programme Planning should involve local people
2. Factors influencing participation of rural people
3. Characteristic of good leader
4. Objective of rapport building
5. Constraints in implementation of programme at grass root level
6. Importance of project evaluation

Ques. 6: Difference Between ANY TWO of the following

1. Monitoring and Evaluation
2. Plan of Work and Calendar of Work
3. External Evaluation and Internal Evaluation
4. Autocratic Leader and Democratic Leader

Ques. 7: Answer in Detail ANY ONE of the following

1. Steps in Programme Planning
2. PRA Techniques
3. Five Year Plan

www.ingramcontent.com/pod-product-compliance
Ingram Content Group UK Ltd.
Pitfield, Milton Keynes, MK11 3LW, UK
UKHW020206290926
15470UKWH00001B/36

9 789351 242062